Modern Misconceptions

By Steve Preston

1st Edition

© Copyright 2017, Steve Preston
All rights reserved.
No part of this book may be reproduced, stored in a retrieval system, or transmitted by any means, electronic, mechanical, photocopying, recording, or otherwise, without written permission from the author.

Table of Contents

MODERN MISCONCEPTIONS 1
TABLE OF CONTENTS 3
INTRODUCTION 5

Biological Misconceptions 16

MEN HAVE MORE TEETH THEORY 17
BIOLOGICAL MAKEUP THEORY 18
LEECH THEORY 21
ARYANISM THEORY 22
POLYGENISM THEORY 27
CRANIOMETRY THEORY 31
PHRENOLOGY THEORY 34
MOLEOSCOPY THEORY 39
PHYSIOGNOMY THEORY 45
BORN WITH HOLE IN HEAD THEORY 48
LIGHTNING BIOLOGY 51
HUMAN COMBUSTION THEORY 55
BABY KILLING THEORY 59
TREPANATION THEORY 62
HEROINE THEORY 66
RADIUM THEORY 70

Anthropologic Misconceptions 77

SPONTANEOUS EVOLUTION HYPOTHESIS 78
STONE BONE THEORY 83
ADVANCED MESOZOIC PEOPLE 87
EXPERIMENTATION THEORY 89
CONTINUED BIO-EXPERIMENTS 99
NEANDERTHAL THEORY 107
CHROMOSOME THEORY 117

NO BONE THEORY 119
STONE AGE THEORY 123

Physics Misconceptions 132
BIG BANG ANOMALIES 133
BODE'S BOGUS LAW 136
THE UNIVERSE ANOMALY 140

Earth Misconceptions 142
PLATE TECTONIC THEORY 143
UPLIFTED MOUNTAINS THEORY 147
MISSING CONTINENT THEORY 156
NUCLEAR TIMING DISASTER 159
FAST EARTH THEORY 173
EARTH STABILITY HYPOTHESIS 180
FLAT EARTH THEORY 186
GLOBAL WARMING HYPOTHESIS 191
ABOUT THE AUTHOR 209

Introduction

This book looks at what can comically be called science. I take that back there isn't anything comical about it. Those calling themselves scientists and establishing theory after theory of nonsense so that they can write papers or textbooks, or try to influence people concerning what they would like to be real rather than any sort of honest scientific test or analysis used to broaden the understanding of mankind. Unfortunately, whenever the word scientist is pushed, people get all mushy and start believing crap. I think it is time to hold what can be called "agreement" scientists, who gather together under a banner of consensus, or the "vain" scientists, who think what they would like to be true is all that is necessary to call something a valid theory; accountable. Take away scientist from their name and take away their "describing theories" privileges. Here are just a few of the items I'm talking about.

***Women have fewer teeth Theory**- described by Aristotle and other Greeks.*

***Trepanation Brain Release Theory**- For thousands of years, scientists and doctors drilled into skulls to release bad stuff.*

***Heroin cure all Theory** -helping society with Heroine soap, heroine toothpaste, Dementia reversal medication the cure for coughs, and all the rest to cure just about everything.*

***Radium cure all Theory**- As Heroine people began dying, scientists grabbed on to Radium and they helped society with* radium chairs and testicle warmers and all the rest to cure just about everything.

Moleoscopy Theory- Agreement scientists channeled their inner knowledge and established a well-received theory. Mole placement somehow would affect how people lived, their success, and their nature.

Flat Earth Theory-While it never was a theory of sailors, textbooks continue to claim Columbus' bravery by fighting off concerns for falling off the edge of the earth.

Global Warming Hypothesis- Agreement scientists that knew nothing about how the atmosphere absorbs energy laughingly came up with an idea that CO_2 could heat the air even when CO_2 emits 10.6- micron energy that cannot be absorbed in the nitrogen based atmosphere. When shown world heating follows sunspot action and sun distance, they just pointed to coal and made over 100 different earth warming prediction programs that have all failed again and again.

Dumb Neanderthal Theory- Even knowing Neanderthal and Cro-Magnon brains were significantly larger than modern humans, vain scientist continue to indicate Neanderthal grunted and clubs their prospective wives.

Humans didn't live with Dinosaurs Theory-Even when hundreds of fossilized modern human footprints integrated with dinosaur footprints, petrified bones, and manufactured artifacts are found all over the world. Vain scientists continue to write textbooks about Australopithecus becoming Homo-Habilis; becoming homo-erectus; becoming Neanderthal etc.

Plate Tectonic Mountain Building Theory- Even knowing the mountain range from Antarctica, thru South and North

America, and around to Asia that surrounds the massive hole that is the Pacific Ocean, is much more than ½ way around the earth. Of course, this would force mountain making plates to be thrust in many directions simultaneously which is impossible but this absurdity is still being put in textbooks.

Nuclear Decay Timing Theory- Even after hundreds of tests showing a 5000% variance in decay determined dates and great knowledge about why the errors cannot be controlled, agreement scientists continue to push the Cretaceous Extinction as 65 million years ago [by nuclear decay] instead of 120 thousand years verified by the combination of Ice core temperatures, O_{18} variance, Pacific Ocean Hotspot tracking, and Atlantic Ocean spreading Paleo-magnetic tracking.

Uncontrolled Evolution Theory- Ignoring the Law of Entropy completely some jerk took Darwin's observations and turned them into a massive "miscalculation".

Land-bridge Theory- While every haplographic DNA specialist will tell you DNA mutations of American s are not found in Asia, some still describe people walking thousands of miles, on ice, to get to Mexico. The absurdity, just made me giggle a little.

Bode's Law- This stupid thing has the unstable planet positioning rock solid so many solar system questions can't be answered. Bode's Law never even approximated the positions of the planets; requires the orbits to stay constant; disregards obvious shifts like Venus and Mars; and comes to a conclusion with a completely absurd mathematical model. Oddly, it is still taught in schools

Train Speed Hypothesis- In America Vain scientists determined that if a train went faster than 21 miles per hour the people on the train would not be able to breathe so no *attempts were made to speed up the trains until the middle of the 19th century.*

No Rocks in Space Theory- *Until the 19th century, there was vain by astronomers that no rocks were in space, so no rocks could fall from space. The strange part of this one is that many had witnessed meteorites falling from the sky but the science community of the 18th century would not accept this abomination.*

I could go on with bogus theories of Phrenology, Craniometry, Polygenism, Bode's Law, electricity invention, and the development of the country India theories and the others, but I think you get the picture. Science stinks. I don't mean real science, but what we find today as *"respected, vain, scientists"* come up with the most bizarro, fictitious, brain melting, silly things. They write papers and others do the unthinkable. Because the writer is either well liked or has a level of fame, others come along and AGREE with the absurdities. When I started looking at this stuff, I found it over and over again and, this is going to get you mad. These great writers of the absurd are the ones that make the text books used in schools as school board members do seem to even read the things. Before we get too deep into all of this, let me provide you with a couple of my own example. The first is the Eye Theory. I think these will help you understand what some are doing to us as we look at biological sciences.

Theory of the Eye

When investigating the number of eyes that a spider had, it was discovered that it had eight eyes which matched the number of legs that it had. Scientists were thrilled and tested a number of other animals, Man had 2 eyes and 2 legs as did a monkey, kangaroo, and on and on went the list. Again, the investigation continued. "How many eyes does a starfish have?" was questioned. It was discovered that the starfish had 5 legs and 5 corresponding eyes. I was just like the previous experiments. A blind worm was tested next. To their amazement, the worm had no eyes which corresponded to his not having legs.

The emboldened scientists proposed the "Leg to Eye Count Theory" which stated that the "quantity of legs is proportional to eye quantity". They finally tested the theory against the bee, but everything fell apart. "Six legs must mean 6 eyes", they insisted. Well, the bee actually has 5 eyes, 2 regular eyes and 3 light sensing eyes on top of its head. To make the theory work, all the scientists had to do was cut off one of its legs. The image below left shows the 5 eyes of a bee which now corresponded to number of legs.

If a theory doesn't work, cut off one of the legs.

My next example is in the area of Etymology

Origin of the term "PEON" Theory

Researchers began looking into the beginnings of the word "PEON' a lowly servant. It was found that during the Victorian Age, indoor plumbing was not considered. Instead; public buildings made a simple yellow line to instruct patrons where to relieve themselves on the floor. The Louvre, in Paris, still has one of these "locators". Homes in the cities used a different method. Under beds were located wide mouthed chamber pots as shown above right. As these were filled, contents were dumped out on the streets. Besides an unpleasant odor along apartment buildings, it was described as the gentlemanly thing for a man to walk on the outside [where feces and urine piled up], and a gentleman spreading his cloak over an area for a lady to walk across had little to do with water. Anyway, the lower level servants actually had the unpleasant job of pouring the chamber pot and invariably, some of the effluent "pee" would get on them. A sign of a lowly servant became pee-on them and a new word was born. My next example is associated with Human Anatomy

Vocal Cord Flexibility Theory

If you have ever noticed, men have a deeper voice than women. This is because the vocal cordage in females is lighter than that inside a man's neck. The following image left is of a female and to the right is the much heavier male vocal cord set. Just like a violin string, the light cord makes a high tone. Besides this piece of news, we find that women say many more words than men. This is because of the vocal cord thickness. Like a violin a female's cord can be sounded with a light touch, while the man's cord, like that on a base fiddle requires much more energy to pluck it or to say a word. Therefore, men are simply unable to provide the energy that would be needed to activate the massive vocal cords as many times as a woman does her tiny cords.

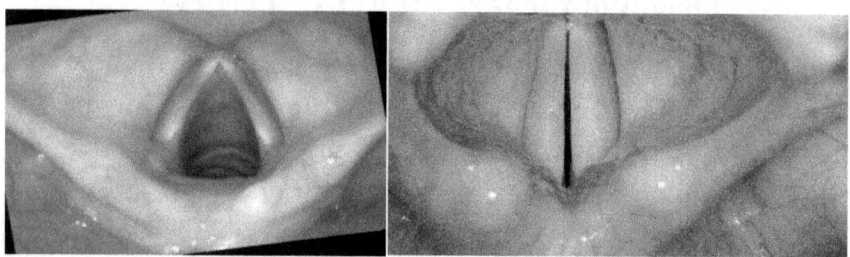

Of course, there is no "eye/leg" count theory, the word peon come from the word "pee", and men can say just as many

words as women if they knew what to say; but these illustrate that, many times, science is quick to use limited testing to prove a theory and will do just about anything to modify facts so that they conform with pre-established doctrine. Many times, this sort of theory survives for many years. The reason is something called stupidity.

Stupidity-When I say stupidity, it is not the same as you typically think in that there is a special different between stupidity and banditry. Both of these theory controllers are suggested in the previous description. One of the reasons that debunked theories can be sustained so long is this stupidity thing.

A stupid person [or scientist] is a person who causes losses to another person or to a group of persons while he derives no gain or incurs losses himself.

I'm not talking about actions that resulted in on person's loss and another one's gain, like a bandit taking our stuff. I'm not talking about someone gaining because of his or her efforts, like some great discovery or winning a race. I'm not talking about 2 or more people losing because of unforeseen events like a flood or something. I'm not talking about 2 people winning because of some major effort on their part. I'm talking about instances where both parties don't gain. The big bang theory would be one of those things. No one win's here. The people presenting the theory lose in that they are so focused on it being successful that they cannot see the limitations of the thing reinforce their conviction by the complete buy in of the "scientists" bringing forth the theory. Of course, a stupid scientist could very well develop a theory, write a paper and publish the "well tested" theory. Rule of thumb if he says he is a scientist, the theory probably has many holes in it.

Sometimes, people need to believe in something even if it's wrong and all of us have substantial portions of our days filled with stupidity. OK! You may not, but I certainly have had my fill. Our daily life is mostly, made of cases in which we lose money, time, energy, appetite, cheerfulness, or good health because of the improbable action of some person's actions who had nothing to gain and indeed gained nothing from causing us embarrassment, difficulties, or harm. Many times, there is no other person and we are doing things to ourselves. Nobody knows, understands or can possibly explain why that person did what he did. In fact, there is no explanation. Better yet, there is only one explanation: the person in question is stupid. Most theories, unfortunately, fall into this region and that is why I'm writing this stupid book. The first part of the eye/leg example would fall into the stupidity region. There were no winners. Children who would be taught this theory would have been the losers, as it would have stifled their learning. To make the eye/leg theory so horrible is the second phase, where details of fact are integrated with fictitious elements [pulling off the leg]. This is where we get into scientific banditry.

Banditry-This is the second most common reason for inappropriate theories to be pushed on us. While slightly different than stupidity, it is far worse. With this type of scientist, we fall in the stupid category, but the theorist may get a gain out of his endeavor, sort of an incentive to make his theory work even if flaws are noted. He may get fame, the Nobel Peace Prize, money, and other things. That whole global warming thing would be an example of the kind of self-serving event that we need to be cautious about.

The perfect bandit is one who, with his actions, causes to other individuals losses equal to or greater than his gains.

The crudest type of banditry is theft. A person who robs you of 100 dollars without causing you an extra loss or harm is a perfect bandit: you lose 100 dollars, he gains 100 dollars. The bandit wants a plus on his account. Since he is not intelligent enough to devise ways of obtaining the plus as well as providing you with a plus, he will produce his plus by causing a minus to appear on your account. All this is bad, but it is rational and if you are rational you can predict it. You can foresee a bandit's actions, his nasty maneuvers and ugly aspirations and often can build up your defenses. This is one of the things we must be cautious about when viewing theories.

Helplessness-That helpless people are those who do not normally recognize how dangerous stupid and/or bandit scientists are. The truly amazing fact, however, is that also intelligent scientists and bandit scientists often fail to recognize the power to damage the helpless. While one is tempted to believe that a stupid scientist will only do harm to himself, this is confusing stupidity with helplessness. One major caution here is that often times stupid individuals are duped into helping the Bandit scientists to affect the helpless.

Helplessness might have reared its ugly head as people have been almost forced to believe that our entire Earth is being turned into a global heater because we burn fossil fuels. While a tiny amount of fact is mixed with tons of mumbo-jumbo, the helpless are forced into believing because hundreds of stupid or other helpless witnesses to the impending doom, fuel the fire so that a very few people can gain from what some tout as a Theory.

***Rule of thumb:** If someone tells you about a theory don't believe the majority of it. You probably can find a little bit of truth, but don't bet your life on it.*

Misnamed Theory-A definition for theory is a hypothesis that has been scientifically tested to show compliance with the initial hypothesis. A corollary to this is; if evidence does not support the theory, it SHOULD be pushed back into the hypothesis level.

With that in mind, we must understand that most of the so-called theories running around should have their name changed to hypothesis. To not change the name is not only destructive, it's criminal.

Here is the scary thing. Our "quasi-science", accepted as FACT, relies on people not asking the hard questions. In this book, we will ask some of the questions that should be provided in classrooms. Just because a scientist says it, it is not gospel. Things like the Big Bang Theory, the creationists theory, the theory of the atom, Newton's theory of gravity, and the plate tectonic theory of mountain building, have all been debunked completely, but still they are taught in schools, addressed in science symposium, and praised in documentaries. Any anomalies to the "tightknit science club" proclamations are tied in a bow and thrown in a hole. We are going to dig some of them out right now.

This book will examine some of the more outrageous "theories" and, hopefully, it will open your eyes to a different world; a world that is not covered up by mistakes made by some scientist trying to get a Nobel Prize. While I'm not what you would consider a scientist, I was, by trade, an Electro-Optics Engineer who developed a number of unique patents and quite a few science related books, showing I have been focused on a number of the sciences for a long time. I will try to piece together a more reasonable answer for the many mistakes of those who would be called scientists.

Hopefully, you will enjoy this overview and learn not to completely trust what you were told over and over to be the truth. The ignorance of truth stretches from Medicine, Physics, Meteorology, Physiology, Astronomy, Engineering, and just about all sciences and many times we never even find out about it. Once the "error" is found out, text books keep the vain lies going for many years as governments, and large communities of powerful entities do not want to disrupt the truth they had built. The one thing that seems to be fairly consistent in this "Vain Science" world is that the initiators want to "save the world from the world". To do this the initiators come to a conclusion and look for a cause to support and agenda. They will go to any lengths to make their hypothesis work. Let's first look at various Biological sciences and pretty strange theories.

Biological Misconceptions

Men Have More Teeth Theory

Simple Theory—Men have more teeth because they are better than women.

The ancient Greeks were known for enlightenment. They did rediscover many scientific truisms, but many times they simply made things up and it was considered science. Here is one of those "enlightened" moments. On one occasion, several Greek philosophers, including Aristotle, had gathered to discuss how many teeth a woman had. Since men had thirty-two teeth, and women were ``inferior'', they all had decided that the latter had fewer --- twenty-eight --- teeth. Not one of them had suggested that they might open a woman's mouth and count the teeth, so the count remained a scientific fact. By the Roman times, some woman counted her own teeth or the scientists started finding ancient apes with the same teeth as Greek men and the theory was reluctantly discarded.

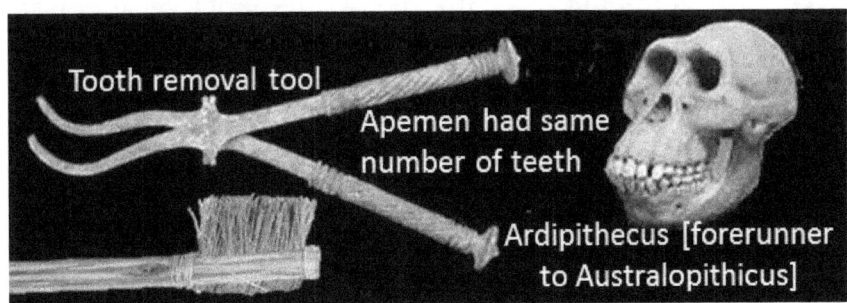
Tooth removal tool
Apemen had same number of teeth
Ardipithecus [forerunner to Australopithicus]

Biological Makeup Theory

For this section, we will go back to the time of the Greeks, to look at a "agreement" "known characteristic of our biology" that would govern medicine for over a thousand years. This was initiated by the father of medicine, Hippocrates.

Greek physician Hippocrates (460 BC to 370 BC) is often credited with developing the theory of the four humors. In this well tested theory, it was established that the body was run by these humors; blood, yellow bile, black bile, and phlegm, and their influence on the body and its emotions was substantial. If you had too much phlegm, for instance, your body would have some issue or your character would be manifested in a negative way. Too much blood could be corrected by bleeding or later by adding leeches to a wound to correct the destabilizing humor. As shown below humors worked in "agreement" with the 4 elements of the Earth, Fire, Air, and Water [whatever that meant.].

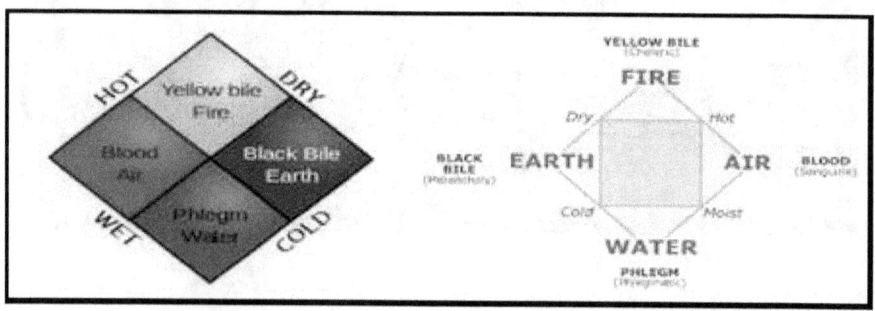

As you would expect from it well tested correctness, this science remained by "agreement" right up to modern times. In this science, it was believed that the right balance of these four humors made a person healthy but an excess or decrease in any one of these would cause illness. Because of this belief, treatments of sickness would include bloodletting, purges, and emetics. Almost all medieval medical *treatments* were designed to *cure* by restoring the natural balance of the *four humors*. Doctors knew that a person's health and personality were dictated by the 4 humors as shown below.

Sanguine personality- *blood controlled-* happy, generous and optimistic, but irresponsible personality.

Choleric personality- *yellow bile controlled-* violent and short tempered, but ambitious.

Phlegmatic personality- *Phlegm controlled-* Sluggish, pallid, and cowardly

Melancholic personality- *Black bile controlled-* Introspective and sentimental

Other definitions are captured in the following graphic.

Humor	Season	Element	Organ	Qualities	Ancient name	Modern	Ancient characteristics
Blood	spring	air	liver	warm & moist	sanguine	artisan	courageous, hopeful, amorous
Yellow bile	summer	fire	spleen	warm & dry	choleric	idealist	easily angered, bad tempered
Black bile	autumn	earth	gall bladder	cold & dry	melancholic	guardian	despondent, sleepless, irritable
Phlegm	winter	water	brain/lungs	cold & moist	phlegmatic	rational	calm, unemotional

Occasionally a mixture of herbs would be used to restore the balance. The humors were also applied to foods – for example wine was choleric (yellow bile). This classification still exists today to some extent, as we refer to some foods as "hot" and others as "dry". To cure melancholy and mental illness, the treatments included the manufacture of blisters to regulate humors and forcing vomiting. Vomit machines were used successfully [just ask the "agreement" scientists].

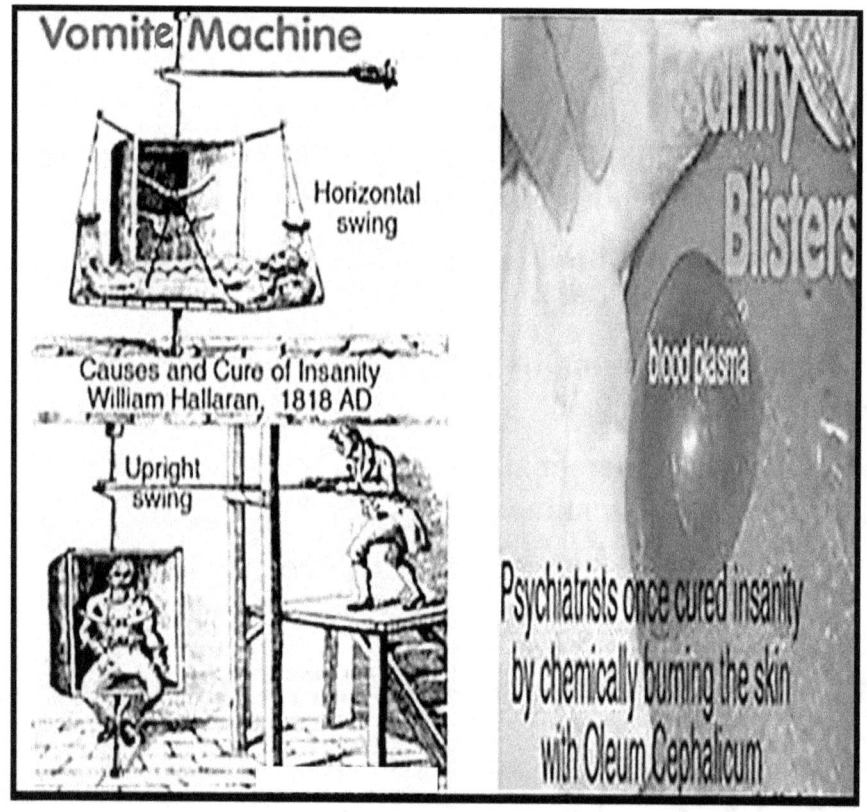

The concept of humors was not replaced until 1858 when Rudolf Virchow published theories of cellular pathology. To make the world a better place, we also have "agreement" scientist that used Anthropology to save the world from de-evolution. If you think I am not a fan of "agreement" science over real science, you would be absolutely right. The following sections may help you feel the same way.

Leech Theory

This theory stated that leeches could cure disease by bleeding patients.

Medieval doctors were all the rage. To cure people, they placed leeches on wounds and made the people bleed. This got the evil out of the body. Some people lived and others die, but the doctors, now called leeches, continued the practice.

Laudanum

Later, the doctor's leeches were replaced with Laudanum [opium]. It started out as the cure for all coughs and ended up as the cure for everything. We will look at that later.

Leeches Today

The fight against Diabetes has been stifled by modern medicine. Circulation is typically lost. As appendages lost circulation, all modern medicine could dictate was to cut off the nasty part and wait until a new section to lose circulation and repeat the process. Someone came along and said, doctors used to use leeches, so let's give it a try. The squirmy things, sucked blood down to diabetic appendages and made the patients well again. Cutting off feet and legs is becoming a thing of the past. Thanks to the medieval doctors and their nasty workers. Well I'll be! This theory was a good one. There are always some good ones around. if you look far enough. Unfortunately, most do not pass the test of time.

Aryanism Theory

In this Biologic Theory, we find that long thin skull people are more evolved than round headed people like Jews and Blacks who aren't even people.

I think a great example of "agreement" science is something called Aryanism. This takes us back to Nazi Germany. While there was some negative distinction about being a black man and the thought that a negroid person was less evolved than normal people; we will look at that sort of "agreement" Evolutionism shortly as this **"science"** was substantially different. This was the desperate attempt to push out or eliminate the round headed Jewish populations to save the country of Germany and eventually save the entire world. This public service was initially established by head shape, but later additional criteria allowed for expanded exterminations.

Georges Vacher de Lapouge- [1890] - He came up with a somewhat new concept in his book "*The Aryan and his social role*" claiming the superiority of the Aryan race and divided humanity into various, hierarchized, different "races", spanning from the "*Aryan white race, dolichocephalic [long thin skull]*", to the "*brachycephalic [round skull] mediocre and inert race*", best represented by the "Jew."

Dolichocephalic simply means having a long thin head while brachycephalic is a round headed, low class person. This included the "*Homo Europaeus* (non-Aryan Teutonic and Protestant), the "*Homo Alpinus*" (French Auvergnat and Turkish), and finally the "*Homo Mediterraneus*" (Spanish and Italians.). "Homo Africanus" (all black people) were not even considered people. Vacher de Lapouge became one of the leading inspirations of Nazi anti-Semitism and Nazi ideology. Some even suggest he sparked the fire for the extermination of the Jewish Poles. Here is an excerpt of his writing.

In the next century, people will be slaughtered by the millions for the sake of one or two degrees on the cephalic index. That will be the sign, replacing the biblical Sabbath and the linguistic affinities that are now the markers of nationality, only it will not have anything to do, as it does today, with questions of moving frontiers a few kilometers; the superior races will substitute themselves by force for the human groups retarded in evolution, and the last sentimentalists will witness the copious exterminations of entire peoples.

William Z. Ripley [1900] - He came up with almost identical characterizations with the in *The Races of Europe.*

Anders Retzius [1920]- He used the thin headed Swedes and the Aryan master race and the bad round headed Slavic people as the low-level people, but the story was the same. The German people had no work, no money and were starving. A wheelbarrow full of 100 million-mark banknotes could not buy a loaf of bread at the time. Below are a few of the multiple billion marks notes used in the 1920s for bread.

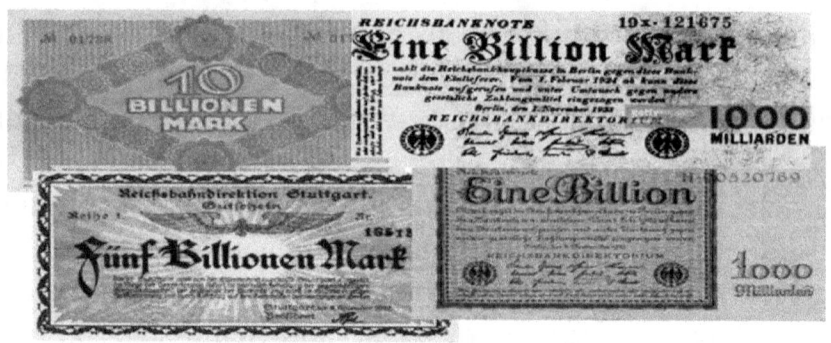

Many Germans were living in shacks after countless homes and farms had been seized by the Jewish control Rothschild/Rockefeller world banks. After Hitler was elected in in 1933, he refused to play ball with the Rockefeller-Rothschild rules and completely thwarting the international banking cartels, many Jewish bankers were killed and, as a result, the Nazi government issued its own currency known as Reich Marchs, which were debt free and uncontrollable by international financial interests. Inflation stopped immediately. Everyone cheered and life got back on track. Hitler noticed that all the people causing the inflation and destruction of his country were Jewish so he talked to his scientists and the ones who wanted to stay alive and prosper told him that Jews were a less evolved group than the Aryan race. Praise and adoration came so another claimed the Aryan features of straight nose, high flat forehead etc. could be used to determine which people should or should not procreate to save Germany and the rest of the world. The "agreement" science community strongly advised Aryan marriage to Aryan Germans to support building a stronger country and money was given for each Aryan child. If a scientist disagreed with Aryanism, he was fired or worse just like those who go against what has been called "human caused global warming in America". Soon almost all scientists accepted and praised the Aryanism assertion just like all "agreement" scientists do as the want to get government grants, write award winning technical papers, get fame, and keep their jobs. Thousands acclaimed the Aryanism science to be great and those not agreeing were in some way against the country. Papers were written, showing how the Aryan race would be able to save the world from destruction of de-evolution. Time went on and Hitler didn't like the Polish people for one reason or another even though they looked Aryan. Scientists set out to establish reasons why the Slavic people should be kept from procreating even with a flat head as the government decided to destroy all 3.5 million of them so the science community pushed the Aryanism to include all Jews as there was a baseness that could not be allowed and the Polish Jews would de-evolve Germany. They had become part of the great awakening to save the earth. Some Aryan looking Jewish women were given a reprieve from gas chambers and the like by allowing them to marry non-Jewish Aryan men. Because they evidently believed in Aryanism, they went along and married the men.

Nazi social policies and German "agreement" science placed the improvement of the Aryan race through Aryanism at the center of Nazis ideology and the most effective method to save the world. Those humans were targeted who were identified as "*life unworthy of life*", including but not limited to Jewish people of Poland and everywhere else, slowly began to include criminals, degenerates, dissidents, the feeble-minded, homosexuals, idle, insane, and the weak. For the sake of humanity, they were scheduled for elimination from the chain of heredity. Despite their still looking Aryan, Nazi scientists agreed that extermination of Slavic heritage (Poles, Russians, Ukrainians, etc.) could only be used as slaves or set up for extermination.

Don't get me wrong, there was no "true" scientific reasoning concerning Aryanism, but "agreement" science won out for the sake of saving the world just like it does every time it grabs hold of a society as we find in the notion that human caused global warming is by something as crazy as CO_2 when H_2O causes the most devastating changes in our atmosphere and hemostasis. In this case racial bias and some innate hatred for Jewish people and national pride drove the Aryanism science. For the human caused global warm-ists the drive started out as concern for a warming trend that warped into simple greed and power-lust. Before we get into that subject, let's look at other similar "agreement" science that pushes ideals without science and appears scientific simply because "scientist" help push the idea just like in Nazi Germany. The next "agreement" is with something called Polygenism which was actually a desire to eliminate the de-evolution caused by black people.

Polygenism Theory

Part of this theory goes like this; black people have more flexible toes and Caucasian so they must be more related to Apes and their skin is simple one large freckle caused by too much Bile.

Similar to Aryanism was a "agreement" science called Polygenism. This science was going to save the civilized world from barbarians by establishing classification of people around the world. The "quasi-scientists" simply started out with a "known concept" that black people were somehow less human and all of a sudden, a new "agreement" science was born. Some say Voltaire was the inventor and reason behind the "science".

Voltaire 1734- He wrote a book *"Traité de métaphysique"*. In it we find the following "*Whites, Negroes, and the yellow races are not descended from the same man*". He believed each race had separate origins because they were so racially diverse. To show how diverse he wrote, *"It is a serious question among them whether the Africans are descended from monkeys or whether the monkeys come from them. Our wise men have said that man was created in the image of God. Now here is a lovely image of the Divine Maker: a flat and black nose with little or hardly any intelligence. A time will doubtless come when these animals will know how to* *cultivate the land well, beautify their houses and gardens, and know the paths of the stars: one needs time for everything.* When comparing Caucasians to Negros, Voltaire claimed they are different species: *The Negro race is a species of men different from ours as the breed of spaniels is from that of greyhounds. The mucous membrane, or network, which nature has spread between the muscles and the skin, is white in us and black or copper-colored in them.* Voltaire is shown next left.

1787-Samuel Stanhope Smith – He was an American Presbyterian Minister and author of *Essay on the Causes of Variety of Complexion and Figure in the Human Species* in 1787. Smith projected that Negro pigmentation was *"nothing more than a huge freckle that covered the whole body as a result of an oversupply of bile, which was caused by tropical climates"*. Later we will talk about this bile stuff as many believed during this time another "agreement" presented around 380BC that people were made up of equal parts of blood, yellow bile, black bile, and phlegm so long as all these things stayed regulated, a person would remain healthy and intelligent. *"By Negros having too much bile it messed up their intelligence."* Certainly, a Christian minister would not use "agreement" science to destroy what people thought about black people, but there he was as shown next middle. I know you are thinking the science community could not go along with this untested, undeserving "theory", but you would be wrong and just as sad as me as George Culver comes along, see next right.

1800-Georges Cuvier- This guy had all the credentials the French naturalist and zoologist. Rather than using some testing and verification he used his understanding of nature and zoological characteristics and the influence of the established scientific Polygenism of Smith. Cuvier "reasoned" there were three distinct races: *the Caucasian (white), Mongolian (yellow) and the Ethiopian (black)*. Each had its place in the world and should be rated by the beauty or ugliness of the skull and quality of their civilizations. Cuvier established the scientific polygenetic understanding of humans.

Caucasians: *"The white race, with oval face, straight hair and nose, to which the civilized people of Europe belong and which appear to us the most beautiful of all, is also superior to others by its genius, courage and activity."*

Negros: *"The Negro race is marked by black complexion, crisped or woolly hair, compressed cranium and a flat nose. The projection of the lower parts of the face, and the thick lips, evidently approximate it to the monkey tribe: the hordes of which it consists have always remained in the most complete state of barbarism."*

1850- Ernst Haeckel – This "agreement" scientist, in his doctrine of evolutionary Polygenism, wrote that *"Negroes have stronger and more freely movable toes than any other* *race which is evidence that Negroes are related to apes because when apes stop climbing in trees they hold on to the trees with their toes"*. He established black people as *"four-handed apes and savages"*, classifying Whites as *"the most civilized."*

Unbelievably he never had a white man and a black man tested for toe motion; it was simply stated by "agreement" and Cuvier never the courage of a group of black people and Smith never tested this bile stuff or had any inclination of what it might be.

Craniometry Theory

In this another extremely important biological construct. In this one, scientists made massively detailed measurements of skulls, nose topography, lip characteristic, and chin placement and size. These table were used to classify individuals. If we could very accurately compare skull shapes we can determine heritage and race to better introduce exclusions.

Craniometry is the study of the shape and form of the human head or skull, sometimes known also as craniology. The difference lies largely in that craniometry implies precise measurement and craniology is simply approximations. If you were wondering what differences were used, below are sketches of the various test characteristics.

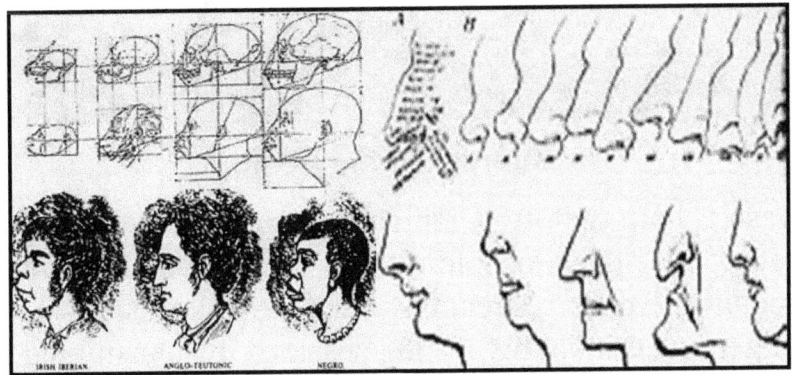

A group of scientists made an obvious observation that people vary considerably in the size and shape of their heads and faces, it is hardly surprising that various scientists attempted to put this factor into a theory to make them famous. The modern and quantitative study of craniology derives essentially from the nineteenth century, when it became widely accepted that evolutionary levels of people could be explored through detailed comparisons of skulls. Actually, this head measurement stuff is part of anthropometry, the quantitative study of the human body which was outlined by Rudolf Martin, professor of anthropology at the University of Zurich around 1900. He provided the systematic basis for craniology and people are still using this absurd classification method. It wasn't long before specialized "tools" to classify people's head were in use around the world. A few of the contraptions are shown below.

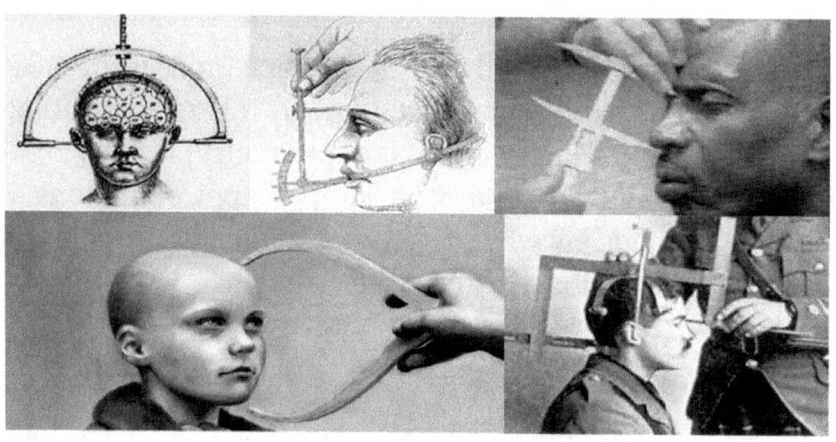

Bumps On the Head

It wasn't long before Craniometry got more esoteric as someone decided people could read an individual's personality, their strengths and weaknesses, hopes, intelligence, criminality, and desires, by examining the pattern of bumps on their skull. Craniometry became popularized as reading bumps on your head forcing a strange vison of the brain. They say it as lumpy and where the brain had lumps the skull would follow. Therefore, by measuring those bumps, one can infer which parts of the brain are enlarged and therefore which characteristics are dominant.

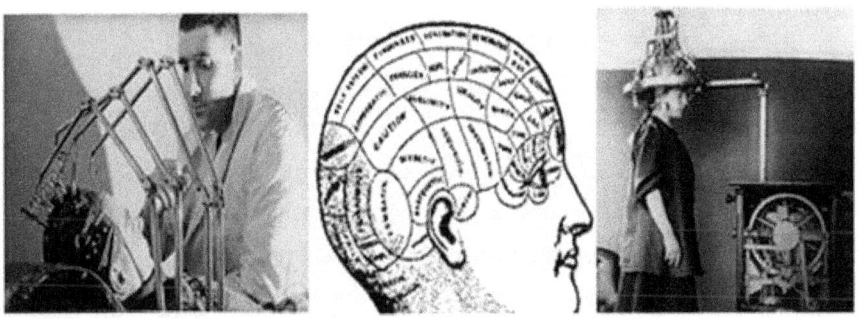

If you think this was and still is stupid, wait until Scientists really get going with an augmentation called phrenology.

Phrenology Theory

> *This was an extinction of absurdity in biology and was considered good medicine for quite a while. Going beyond the confines of Craniometry, these biological scientists decided that the distance between the eyes determines how a person will develop and a large jaw makes a person a criminal. From there they could define you any way you wanted for a price.*

This "agreement" science of Phrenology with a bit of Craniometry mixed in was made famous by Cesare Lombroso, the founder of anthropological criminology around 1880. He claimed to be able to scientifically identify links between the nature of a crime and the personality or physical appearance of the offender. This is what some have called "born criminal" unlike some of these guys, Lombroso did do some level of experimentation, but he was so bought into the result he had already introduced, the experimenting did little to modify the theories. He concluded that skull and facial features were clues to genetic criminality, and that these features could be measured with craniometers, goniometers, and calipers, as shown below, with the results developed into quantitative research.

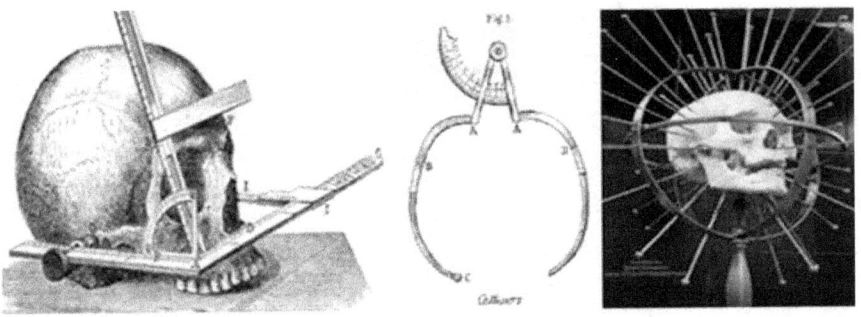

A few of the 14 identified traits of a criminal included large jaws, forward projection of jaw, low sloping forehead; high cheekbones, flattened or upturned nose; handle-shaped ears; hawk-like noses or fleshy lips; hard <u>shifty eyes</u>; scanty beard or baldness; insensitivity to pain; long arms, and so on. Here is a news article of how this nonsensical science would help mankind classify criminals.

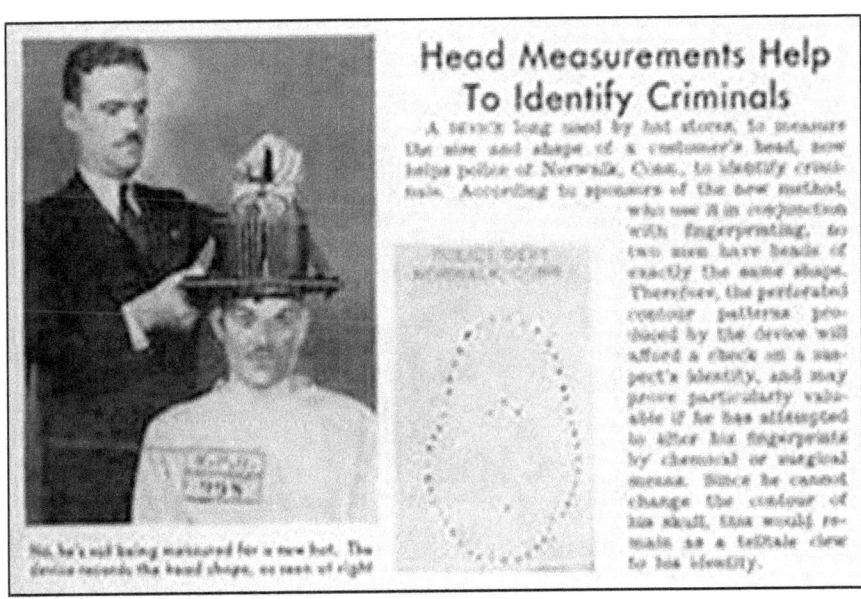

Massive devices full of test points assured the testing would be thorough and unbiased. This even extended to electronic measurements as shown below right.

Facial Angle Testing-Pieter Camper [1750] had distinguished both as an artist and as an anatomist, published some lectures containing an account of his craniometrical methods. These laid the foundation of all subsequent work. He was the inventor of the "facial angle", a measure meant to determine intelligence among various species. According to this technique, a "facial angle" was formed by drawing two lines: one horizontally from the nostril to the ear; and the other perpendicularly from the advancing part of the upper jawbone to the most prominent part of the forehead.

While this was also used in the Aryanism attempt to cleanse the world his ideas continued throughout the 19th and into the 20th centuries. Camper claimed that antique statues presented an angle of 90°, Europeans of 80°, Black people of 70° and the orangutan of 58°, thus displaying a hierarchic view of mankind, based on a decadent conception of history as we became more and more de-evolved. The testing became more and more involved as shown.

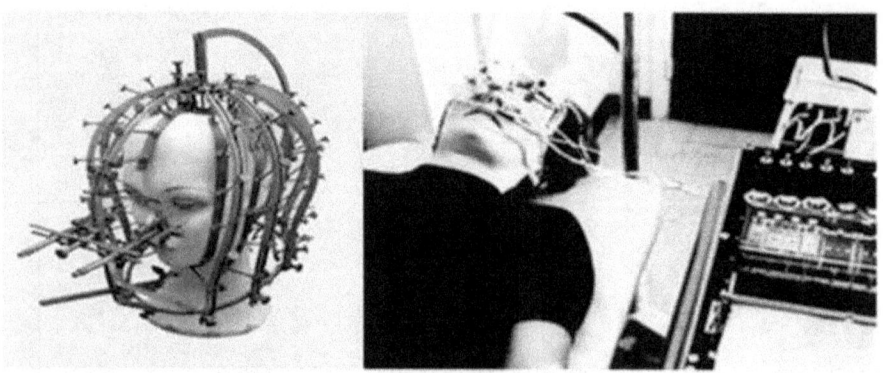

This "agreement" scientific research was continued by many others using many complex devices to assure the testing was accurate and "scientific";

By the middle of the 19th century, phrenology parlors were widespread. Automated phrenology machines came later. The automated machines were composed of numerous spring-loaded probes. The device was placed over the head while the probes would extend to gently touch the scalp, thereby providing a measurement of the topography of the skull. The machine would then calculate the characteristics of the subject based upon this topography and produce an automated reading. One could even test himself using devices shown below.

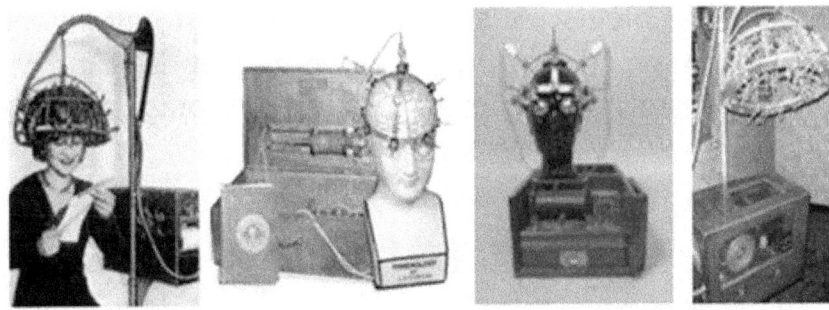

This stupidity continued. In the late 19th and early 20th centuries, the concepts of phrenology became associated with "agreement" scientific ideas in the areas of criminology and evolution that were popular at the time. Craniology and Anthropometry were attempts at identifying evolutionary advancement and criminal tendency according to physical measurements of the skull and face. As with most of these "agreement" sciences, these measurements were used to verify pre-existing social prejudices. To give the testing an air of reasonableness, automatic and robotic tasting machines shot up around the world as we tried to protect ourselves from ourselves. Here are more examples of robotic testing systems that assured the results could be help the world understanding of its inhabitants.

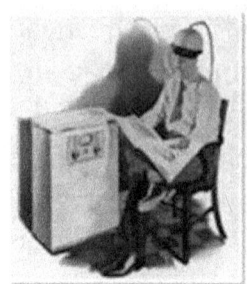  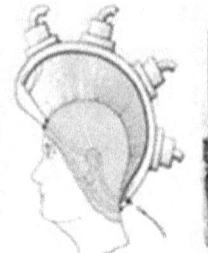 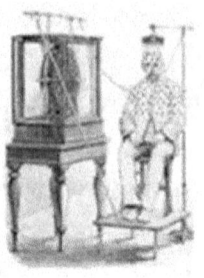

All the Phrenologist scientists were trying to protect the environment of the world by weeding out humans who should not procreate. another test was accomplished with Moleoscopy.

Moleoscopy Theory

> *Unbelievably, "agreement" science claimed a victory in medicine again with something called moleoscopy. The study of moles would determine the nature of a person and active practiced of this began in the 18th and continued as late as the 19th century.*

Doctors would simply check the skin for any moles and help the patient. These classifications could be used to help or hurt getting a job, winning a case in court, or finding the right person to marry. Today many have broken free from "agreement" on this one, but at one time here is the extensive characterization controlled by moles.

Rib Mole—A mole on left rib away from the nipple shows a lazy person

Ankle Mole — On a man this shows a fearful nature. On a woman, a sense of humor, courageous; is willing to share love and worldly possessions with others.

Armpit Mole— Under the left arm indicates the early years are a struggle but ample remuneration, even riches will make the later years very happy. Under the right arm this means constant vigilance for welfare and security must be kept uppermost.

Back Moles- If you have a back mole you must be sure you have all the facts before you enter into negotiation of any enterprise.

Belly Mole– Belly mole people have a tendency to self-indulgence. Avoid overeating and excessive drinking. Keep a rigid check on your economy. Choose a marital partner who has an even, calm temperament and the gift of understanding.

Bosom Mole-- This mole can make the wearer have a quarrelsome nature given to temper; this mole indicates a lazy, sometimes unsteady disposition. Lack of ambition may result in a colorless career. On a man, this area is the chest.

Breast Mole – A mole on the right breast will have, indolence and intemperance that may destroy happiness for self and family. These people need to exert will power and self-discipline so that you can enjoy the love and comfort the children you might have. A mole on the left breast made a person active, energetic, and able to concentrate upon the acquisition of wealth and property.

Nipple Mole-- On a man a nipple mole indicates he is fickle and desirous of many amours. On a woman, it shows she is always striving for social status.

Nose Mole —This mole shows someone who will be a sincere friend. The wearer will achieve success and make an excellent marriage partner. Nose mole people are dedicated to amassing great wealth even though the struggle often seems impossible, however; if the mole is on bridge of nose, the person will become very lustful.

Buttocks Moles—Luckily no one typically sees this mole as is shows a person is not very ambitious and inclined to accept any mode of living, even poverty.

Ear Moles—These are rare, but whoever possesses one may find riches far beyond expectations.

Elbow Moles—This mole shows a tremendous desire to travel and one who is always uncertain. Usually talent in one or more of the arts is associated with this mole. The wearer is capable of earning a fortune in money but rarely has the urge to work for it.

Eye Moles — Poverty overshadows unusual talent in this wearer. If the mole is located on the outside corner of the eye it means an honest, forthright person; one who is reliable but needs love and admiration to offset the struggle for existence.

Eyebrow Moles — Over the right eyebrow, this mole signifies perseverance and a very active life and successful in everything — business, home, and family. Over the left eyebrow or temple, the reverse is threatened. Disappointments will be due to selfishness and indolence. Only with a maximum of effort can poverty be avoided.

Finger Moles — On any finger, this shows dishonesty and an inclination to exaggerate due to inability to face the hardships that must be confronted.

Foot Moles—This shows someone who is inclined to brooding akin to melancholia. The wearer prefers a sedentary life but really needs a balanced amount of activity to remain healthy.

Forehead Moles-- The middle of the forehead mole predicts honors, wealth, love, and a happy, distinguished family. Right and left forehead mole people are similar to eyebrow mole people.

Groin Moles – Despite prosperity, a mole on the right groin shows a propensity for ill health. On the left side, this type of mole shows frailty without much prosperity.

Hand Mole — This shows an abundance of almost everything, health, wealth, and happiness. Usually this person is very talented. Some tattoo on hand moles to see if their lives will change. This writer has no information on outcomes.

Heel Moles— People who are very active mentally and physically have heel moles. This shows the ability to accumulate a fortune if so inclined, but makes enemies who continuously plague and cause petty annoyances.

Hip Mole — A mole on any part of the hips except the buttocks, contentment, fortitude, and ingenuity are the salient attributes that balance an otherwise over-amorous nature.

Instep Mole – Instep moles show quarrelsome nature. They are often sullen and generally have a keen interest in athletics.

Knee Mole — On the right knee we find a friendly, amiable disposition; a great lover, desirous of family and home life. On the left knee, we find extravagant and inconsistent nature, but an excellent business acumen.

Leg Mole— This shows many difficulties during early years but capable of surmounting them by sheer forcefulness. Resources must not be dissipated. Avert any tendency toward indolence.

Cheek Moles--On either cheek this mole shows a serious, studious, almost solemn person. They have a middle-of-the-road point of view on most theories pertaining to living, religion, and politics and consider wealth as not necessary for happiness.

Lip Mole— This person has a benevolent nature, always striving for better conditions. [See below right],

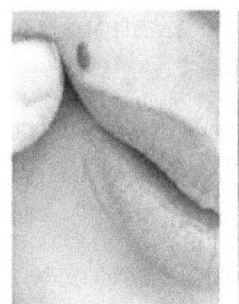 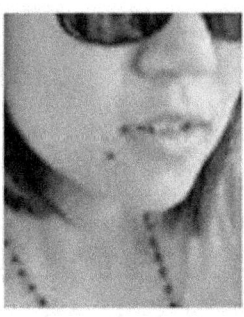 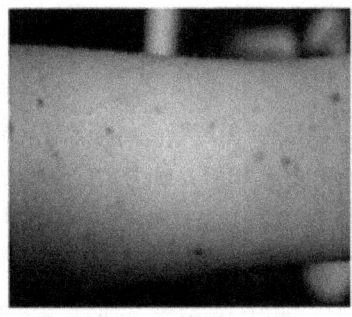

Chin Mole-- Many people have a mole on the chin. Right or left, it designates people with enviable characteristics. These are people with loving, generous dispositions. They are conscientious workers, and love to travel and acquaint themselves with the habits and customs of other peoples in distant countries. They are capable, responsible citizens, willing to accept responsibilities for family and country. [See above middle]

Arm Mole – Arm mole people are courteous, industrious, and have happy conjugal relations. A man may have to fight many battles if the mole is near the elbow. He may also become a widower at an early age. A woman has the same characteristics but her problems are in her occupation. [See above right]

Naval Mole — On a man, he will be very lucky. On a woman, she will desire to have many children.

Neck Mole — This means unexpected good fortune, if mole is on front of the neck; on either side, unreasonableness. On the back shows a need to practice frugality.

Shoulder Mole — Generally, restless, needs to travel in order to be satisfied with home surroundings. A mole on the right shoulder brings prudence, discretion; a faithful marriage partner; very industrious. On the left shoulder, this person is satisfied with any position in life, both occupational and social.

Wrist Mole — Frugal, ingenious, dependable people have wrist moles. Furthermore, on a woman shows she will have only one marriage; on a man, possibly two marriages.

Twin Mole--When there are twin moles, such as a mole on one wrist in a certain spot and an identical one on the other wrist, this is called a Gemini duality. The person possesses a dual nature. This pertains to all dual moles no matter what the location, such as legs, arms, cheeks, and so on. Two moles, side by side, are said to indicate two loves.

For those with more than 2 moles, one could determine he should not have children and pass on the indications of the moleoscopy testing and the world again was saved until Physiognomy scientists came along to change the laughable biological theories into a useful science.

Physiognomy Theory

> *This theory indicates that the wrong plastic surgeon can really mess up your life by making you aggressive and naive.*

This science without science was almost universally praised by ancient Greek mathematician, astronomer, and scientists. While you would think this stuff would not continue, by the 15th century Physiognomy science was widely accepted. English universities taught it. Around 1530, scholastic leaders started with the Greek form 'physiognomy'.

Greek Physiognomy- The first indications of a developed physiognomic theory appear in fifth century BC Athens, with the works of Zopyrus, who was said to be an expert in the art. Here are some details of Physiognomy used in Greece. *After inspecting Socrates, a physiognomist announced that he was given to intemperance, sensuality, and violent bursts of passion—which was so contrary to Socrates's image that his students accused the physiognomist of lying. Socrates put the issue to rest by saying that originally, he was given to all these vices, but had particularly strong self-discipline.*

By the fourth century BC, the philosopher Aristotle made frequent reference to theory and literature concerning the relationship of appearance to character. Aristotle was apparently receptive to such an idea, as evidenced by a passage in his "Prior Analytics"; *It is possible to infer character from features, if it is granted that the body and the soul are changed together by the natural affections: I say "natural", for though perhaps by learning music a man has made some change in his soul, this is not one of those affections natural to us; rather I refer to passions and desires when I speak of natural emotions. If then this were granted and also that for each change there is a corresponding sign, and we could state the affection and sign proper to each kind of animal, we shall be able to infer character from features.* [Remember the dog remark as we go along.]

Chinese physiognomy- or face reading (*mianxiang*) reaches back at least to the Northern Song period. Like the others the desire was to rid the world of inappropriate people.

18th Century Physiognomy-Another principal promoter of physiognomy in modern times was the 18th century Swiss pastor Johann Kaspar Lavater. Lavater's essays on physiognomy were first published in German in 1772 and gained great popularity. These influential essays were translated into French and English. A Belgian by the name of Paul Bouts, expanded the concepts of the early Greeks so he could make the world safer. He combined phrenology with typology (character analysis through body morphology) and graphology (character analysis through handwriting examination). He is noted for calling his three-in-one science Psychognomy. The mole thing didn't seem to track, people, again pushed for science to help them understand people they finally knew moles were just moles so someone began measuring heads outside of the bumps and nose angle. This "science" used facial characteristics to determine nature of a person. Here are some of the findings.

- **Round head** –An intuitive person
- **Triangular head**-An impractical, but quick-thinking person
- **Straight eyebrows**- An alert and active person
- **Arched eyebrows**- A person with a great imagination

- **Large nose-** An aggressive person
- **Narrow lips-** An unemotional person
- **Large earlobes-** An independent person
- **No earlobes-** A person who lacks sense of purpose
- **High forehead-** An intellectual person
- **Narrow forehead-** A great analyst
- **Round eyes-** A naïve person

These evaluations continued to all elements of civility. Here is a sample from James Redfield's book "Comparative Physiognomy" [1852]-*The organ of veneration is situated in the middle of the coronal region, between benevolence [Left below- hump in middle of skull] and firmness [second below- dip in the middle of the skull]. When rapt in devotional feelings, when all outward impressions are unheeded, the eyes are raised by an action neither taught nor acquired. Instinctively they bow the body*

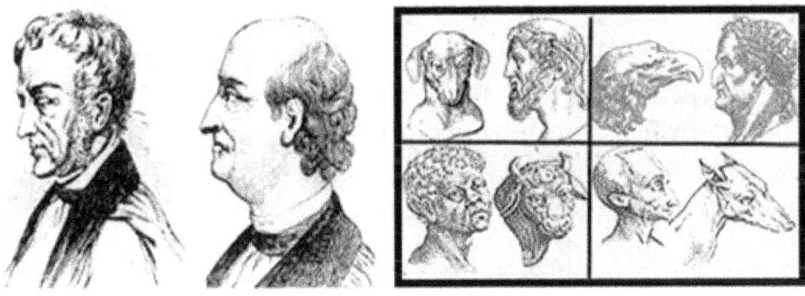

In his book Redfield "*Resemblances between Men and Animals*", he also associated people's character and appearance to animals. [Germans to Lions, Negroes to Elephants and Fishes, Chinamen to Hogs, Yankees to Bears, Jews to Goats] Here are a few of this "agreement" scientific disclosure. [See above right]. One thing that there was not agreement on was why we keep finding skulls with what appear to be bullet holes.

Born with Hole in Head Theory

It seems more and more skulls are being found with tiny entry holes in their heads. The logical theory is they were born with a hole in their heads, so don't even think that ancient humans used rifles in their wars. They were way too stupid.

Can you imagine quasi Scientists and Historians pushing an idea that no gunlike weaponry was available and used in ancient wars. More and more evidence of the use of some type of guns keep popping up and they just say poppy-cock or maybe some believe in the theory presented in the header. The entry hole is something that individuals could live with, but would there have been any damage to the brain? While there is little doubt that nuclear weapons were used ancient wars around 11 thousand and 5500 years ago from the nuclear fallout indication, is it so hard to believe these people may also have used conventional weapons?

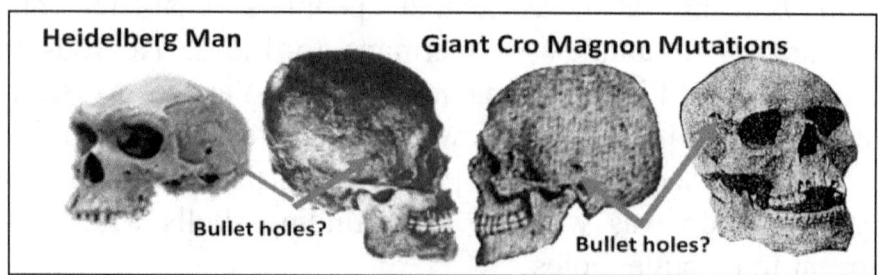

Some scientists started looking for holes and found bullet holes in skulls of Cro-Magnon and even earlier people, pointing to the use of high speed projectile weapons used by the Homo-Capensis humans of the Pleistocene and early Holocene. Here are just a few examples. Even animals appear to have been shot.

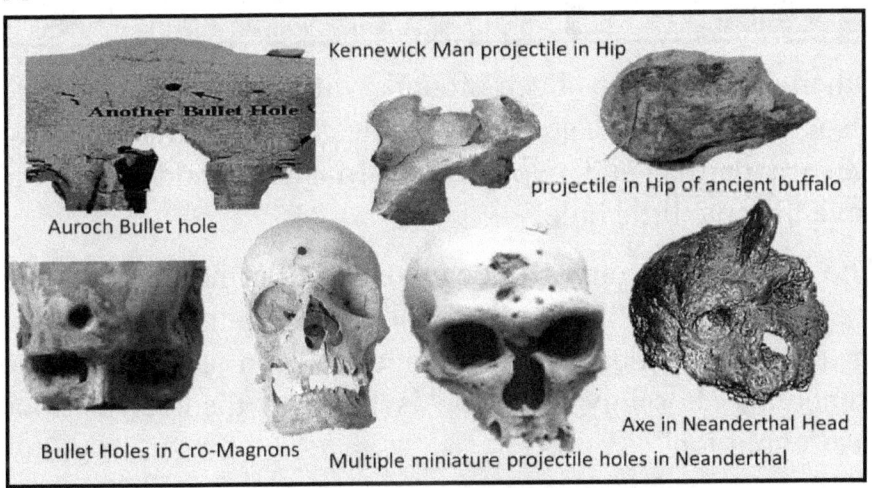

In Russia and extinct Auroch was found with a similar high-speed projectile hole in its forehead which is believed to have been put there during the Pleistocene [See the preceding collage]. A similar hole was found in the same type animal in Zambia and other signs of hostilities. Many had a tiny hole going in and much larger exit destruction. Please notice something that might be bad. Many of the bullet holes appear to be from point blank range given the entrance hole positions as if done by a contract killer.

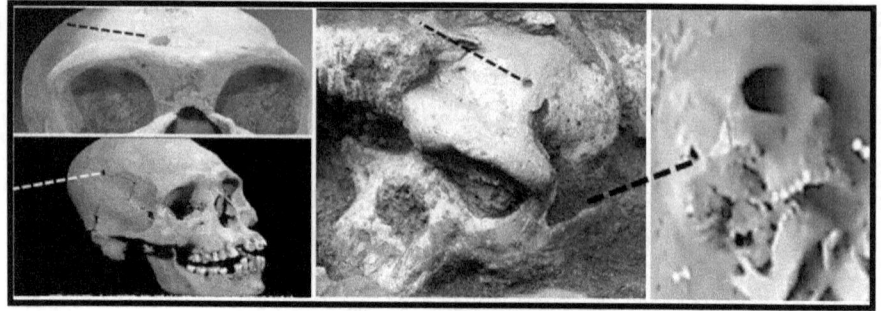

Other skulls have been found with the piculiar holes associated with high speed projectiles and shown above. Two are neanderthal and two are Cro-Magnon/ modern, but all have the cute little hole.

The following group all apear to be bullet holes except for the spear slice in the skull on the upper right. Possible each of these individuals slipped and fell on a steel rod or something. Possibly someone is lying to themselves about our "evolution".

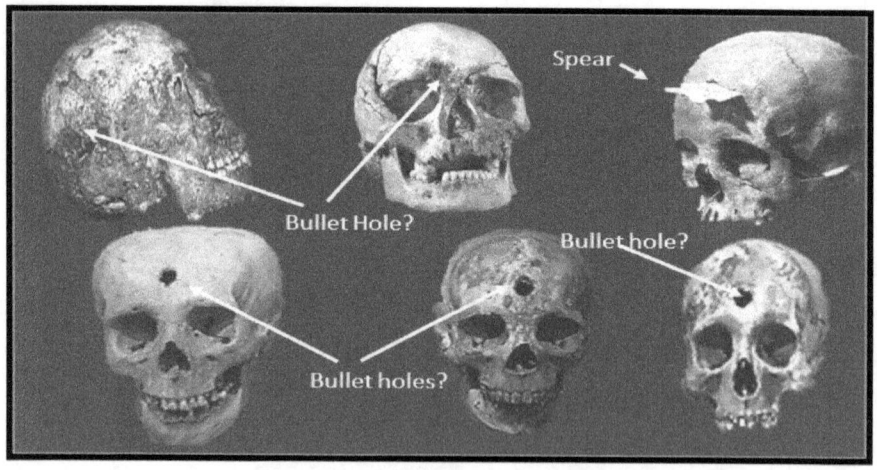

That brings us to another way to die-- with lightning.

Lightning Biology

Can a man's biology attract lightning? If so what are scientists doing about it?

One man we need to talk about now would never have survived a summer shower. He was deathly afraid of even a clear day. The reason Roy Sullivan was so shaken every time a summer shower came along was that he held the dubious record for being hit by lightning more than any other human of the 20th century. Roy's image is show below right, and the strange burn marks common from a lightning strike are shown to the right.

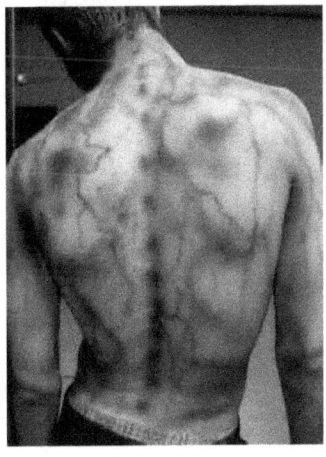

For Roy, it started in **1942**. A lightning bolt hit. He lost only his big toenail. All was good for the next 27 years, but it would not last. In **1969** there was another strike and he lost his eyebrows. In **1970** another hit and his shoulder was seared. In **1972** lightning set his hair on fire. He got smart and luckily kept buckets of water with him at all times. In **1973** a bolt set his hair on fire again and blew off his shoe. He was still alive and he used his water to its full advantage. In **1976** a strike messed up his ankle. In **1977** chest and stomach burns were the result of still another lightning strike. Finally, Roy got out a pistol and killed himself. While you would think scientists would be trying to find out what in a human could attract lightning.

The number one thing to know about lightning strikes is that they are almost always attracted to men.

Possibly lightning is attracted to testosterone, but what is known is that men are struck by lightning four times more often than women. Here are some statistics according to a 1994 study.by the National Severe Storms Laboratory and the National Weather Service. From 1959 until 1994, males account for 84% of lightning fatalities and 82% of injuries. This was from a data base of 3,239 deaths and 9,818 injuries from lightning strikes. For those thinking Mr. Sullivan was just unlucky, let's look at more data.

As with others, a man named Ken Brody, was struck 3 times and survived sometime in the 20th century.

August Hellman of Arkansas, indicated he was struck twice, once while baling hay in an Oklahoma field in 1959 and again while sailing on Savannah Bay in 1977.

A 12-year-old girl named Alice Svensson was struck by lightning twice while taking a shower just before a thunderstorm. After dinner, she went downstairs in the basement to take a shower when it was not raining. Unfortunately, 2 claps of thunder and a massive scream showed lightning had entered the water pipes and through the water to the unsuspecting girl. She got out of the shower and a second bolt hit as her as she was holding the metal shower hose.

Don't even suggest lightning will not strike the same place twice as these targets were moving and still Lightning was able to strike more than once. The following collage shows more of the typical results of strikes, loss of memory, scars, loss of body parts, singed hair, and, of course, death.

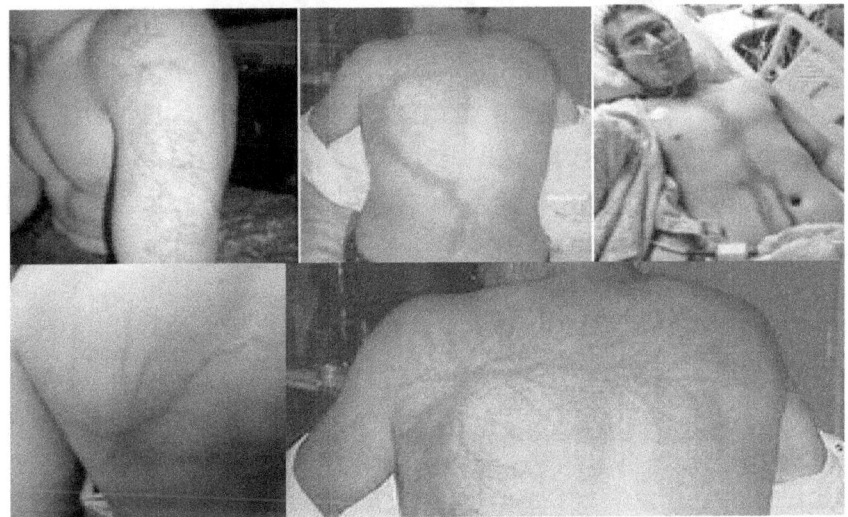

In 2011- Melvin Roberts, shown below, made headlines for being hit 6 times but that isn't the big story. His wife said he's been struck another 4 times. The South Carolina man suffered memory loss, headaches, speech problems and has nerve damage in his hands and leg from his last multi strike event.

A Possible Reason

Finally, let me talk about a girl named Rhonda. She indicated her mom had been hit 3 times, her brother and two of her aunts had been struck. She claimed that it must be a genetic physiological issue and may have something to with her inability to wear a watch as they always stop running if she wears it and her various call phone batteries would never hold much less charge compared to that of her friends. I searched for papers concerning work done to find out how to make us safer or at least let us know propensity to attracting lightning and there seems to be no interest. If we assume, there is a particular DNA combination that allows for the attraction of electricity, the question might be, could one cause a lightning bolt inside their own body?

Human Combustion Theory

Can the Human Biology Cause Spontaneous Combustion?

While being struck by lightning might be a horrible way to die, what if the lightning was already inside you? Most people have heard about spontaneous human combustion and immediately ignored it due to its absurdity even though there have been very few explanations presented for some very bizarre occurrences. Some had even seen pictures like the one to the right that showed the effects of an unexplained phenomenon, but were unmoved. Unfortunately for Jaqueline Fitzsimons, in 1985 it became too real and there were many witnesses of its reality. She just went to college one day, told her friend she wasn't feeling well and that her back felt extremely hot. All of a sudden, her shirt caught on fire and she screamed for help. Many came to her rescue and put the flames out, but not before her hair was a blaze. After 15 days in intensive care, she died. History and science ignore what has been happening in the past. It is now a little harder to ignore. People can catch on fire by some internal means, so if you get heartburn carry antacids and possibly a fire extinguisher.

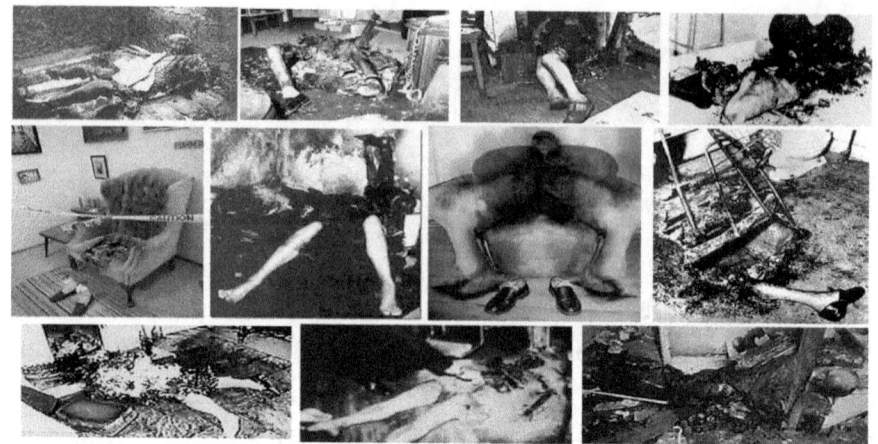

Next is a short list of the more notable instances that have been determined to be spontaneous human combustion in just the past 50 years. This is not an isolated case.

When	What	Where
1951	Mary Reeser burst into flames. Only the corner of the room was burnt.	Florida
1950's	A secretary burst in flames. Her boyfriend was unable to put flames out.	London
1953	Esther Dulin burst into flames. Only she & her chair were consumed.	Los Angeles
1957	Anna Martin burst into flames at home	Penn.
1959	Jack Larber burst into flames in a hospital	Calif.
1959	Billy Peterson burst into flames in a car. The seat was undamaged	Michigan
1964	Fiery death of a man was reportedly like an exploding person	London
1964	Helen Conway's torso was found charred in her home.	Penn.
1964	Olga Worth Stephen burst into flames in her car	Texas
1966	George Turner was found incinerated in a ditch.	Chester
1966	Willem Bruik burst into flames in a car.	Holland
1966	Dr. John Irving Bentley was incinerated at home. His leg was unaffected	Penn.
1967	Robert Bailey burst into flames. Flames were coming from his stomach.	London
1970s	A building Contractor burst into flames while waving to employees.	London
1980	Mr. Blackwood was incinerated but rubber on his walker was not burned.	Wales
1980	Jenna Winchester burst into flames while sitting in her car.	Florida
1982	A woman burst into flames on the street.	Chicago
1982	Jean Saffin burst into flames in her kitchen. Her father remained helpless.	London
1985	Jouqueline Simmons burst into flames at school.	Cheshire
1989	A 27-year-old engineer burst into flames. His stomach and belly were carbonized.	Hungary
1997	John Oconner burst into flames at home. Only his head, upper torso and feet remained.	Ireland
1998	Agnes Philips burst into flames in her car. Flames came from her chest.	Australia
2003	Alexei Rusnac's head was burnt while his clothes were not burned according to "Ananova"	Rumania

What the above list shows is that "Spontaneous Human Combustion" occurrences are all over the place, and people spontaneously combust just about every year. Some years more incidences seem to happen than others and the following general information concerning the symptoms of the mystery are described below.

- Although it doesn't appear that way from my short list, almost eighty percent of the victims are female.
- Most of the victims were overweight and/or alcoholics.
- The body is very badly burned, but the room the body is found is in pretty much intact except for a fine layer of soot.
- Yellow, foul smelling oil is usually surrounding the body.
- The torso, including the chest, abdomen and hips tend to be totally consumed, sparing portions of the extremities and the head - the clothing can also be intact.
- Generally, victims were on their own. Generally, no shouts or screams could be heard. This was not the case with Jaqueline.
- The victim had usually been drinking heavily prior to the death, but that isn't always the case.

No one knows what causes it and no one can predict when it will happen. One possible reason for the deadly occurrence might be vibrational discord.

For years scientists have known the delicate hemostasis of vibrating elements in our bodies. Anger, hatred, fear, and other emotional outbursts not only make people feel hot, their bodies could be changing and given the wrong set of circumstances, the volatile materials could build a chain reaction. Rather than research ways to control out vibrational hemostasis, Scientists typically disregard hundreds of people catching themselves on fire. While SHC can kill, there is another death sentence brewing.

Baby Killing Theory

When is a baby considered to be a living biological entity and worthy of life and when should it be killed?

In 1973, the Supreme Court ruled that ALL unborn children up through age 24 weeks were not people with rights of citizens and they could be killed without recourse. Luckily, many escape the forceps. Here are a few survivors

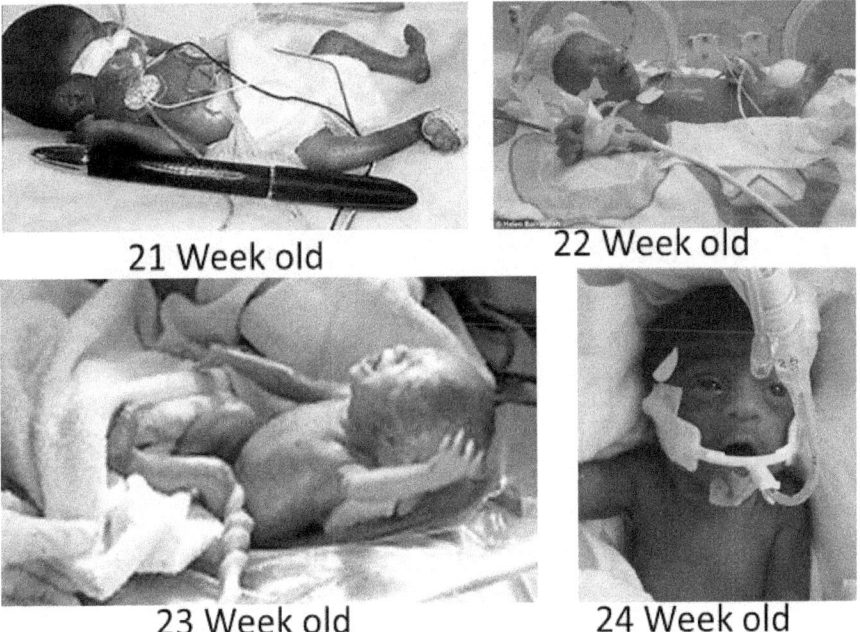

21 Week old   22 Week old

23 Week old   24 Week old

Like many laws this one has substituted biological ascertains over convenience of a woman. When asked when should a baby have a right not to be killed, we find all type sf answers.

Some say when it is not viable—So we know the 21-week deliveries should be held sacred.

Some say if the heart cannot beat on its own------This would certainly allow us to kill those requiring pacemakers to keep their hearts going.

Some say if a fetus cannot breathe on its own- That would not survive well for patients on ventilators.

Some say when someone has useful brain function and then we can kill those in comas who could survive.

With fetus' sucking their thumbs, yawning, moving on their own, and even getting hiccups after 4 months, it seems the killing limit should not be less than 16 weeks.

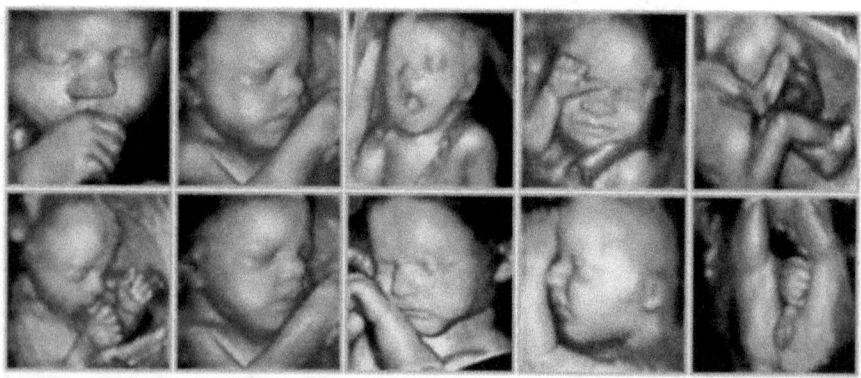

Soul Entry- This is more difficult as one cannot be sure when a soul enters a living body. For a tiny fetus human to move around as his body reacts to this world well before the time of birth, it HAD TO HAVE OBTAINED A SOUL. Sorry, but I have to say this. The 5-month-old fetus shown above is targeted for destruction in this strange world we live in where no morals are considered acceptable. This "abortive destruction" will cause the soul to leave the body as he or she is murdered

Some say less than 9 months from inception--- others say let's make that 18 years after inception.

Some say for the protection of a girl who will have a hard time keeping a baby--- Sorry; no sympathy for sexual activity out of wedlock and not taking responsibility from me., in fact killing anyone because they make you feel uncomfortable should be against the law.

Some say in cases of Rape; as if the baby somehow had part in the rape and should be killed for it.

Some say to save the life of a woman who is in imminent danger from deliver. This should be considered always. This is perhaps 5 or 6 killings a year I would imagine.

If killing babies is simply for population control let it be addressed that way and see if Americans want that level of barbarism be the law of the land. Speaking of barbaric practices done in the name of scientific cure, we have to look at trepanation.

Trepanation Theory

The Science of Trepanation is simple drill a hole to the brain so it can be free to expel anything bad.

Rather than having a documented reason tested by science of any kind. This has been practiced for over 30 thousand years. Neanderthal skulls have been found that have had a hole drilled into the skull for so medical relief. Like the other elements discussed in this book, trepanation was practiced worldwide and accepted as proper medical treatment for a number of issues. We know that trepanation was done for epilepsy, infantile convulsions, headache and various cerebral diseases believed to be caused by confined demons to whom the hole gave a ready method of escape. While you would think this barbaric process was only done in very ancient days. We can find records and images showing the process being done in the 13th through the 21st centuries as shown below.

Sometimes with special drills with stops to insure the brain is not destroyed, but most of the time, simply by stopping the drilling at the appropriate time. I must admit there is a variant of trepanation called craniotomy used today, where a neurosurgeon creates a burr hole to relieve the pressure of a hematoma or brain bleed. That variant is not what we are talking about here. In ancient Peru, unusual long headed skulls different from modern skulls were found with trepanned holes to fix brains as shown below.

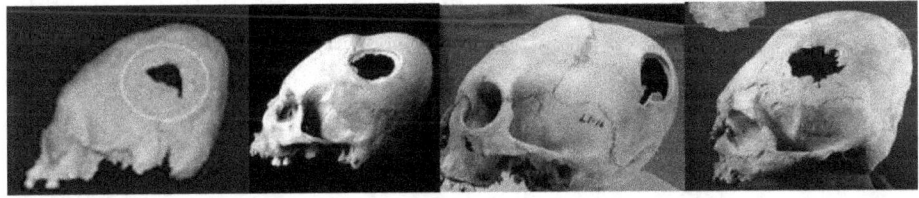

As drills were in short supply in the very ancient day, straight blade knives were used to saw through bone and matter as shown below.

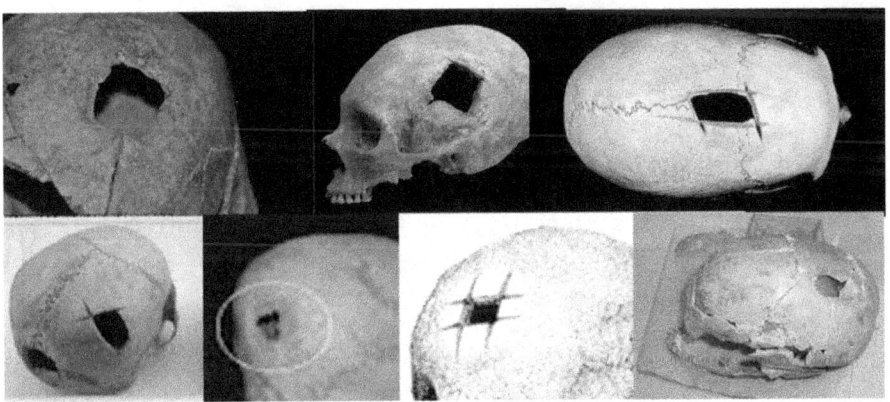

From many skulls found, it was very hard to get out of the skull what the "doctor" desired so a larger and larger hole was established. Some of these holes were enormous and evidence shows the patients did survive the process. The group following is a small sampling of the tremendous steps taken to relieve patients.

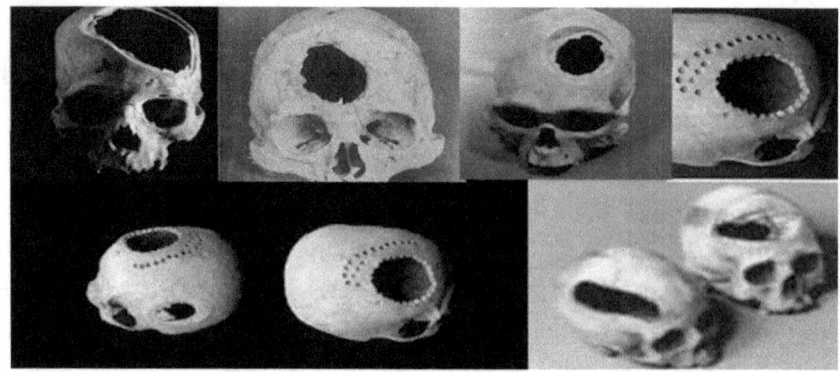

Many times, a small hole would work or many holes all over the head. Some of the skulls shown in the following group showed that the patient went back to the "doctor" many times for more holes.

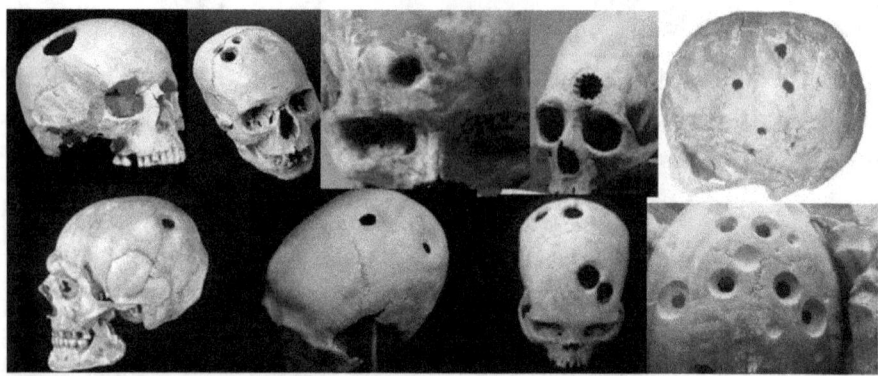

Some doctors got creative to make oval holes. Possibly this was to present a desired look after the surgery or a doctor's signature. [See the following group]

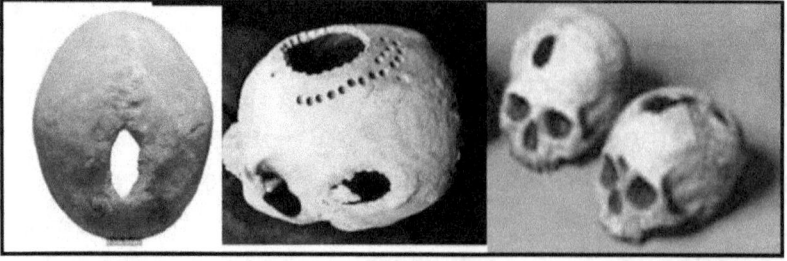

While some may have been told there would be no scaring or noticeable indication of the procedure, pictures of people who had these processes done show that the loss of skull is a noticeable characterization of those being put under the knife.

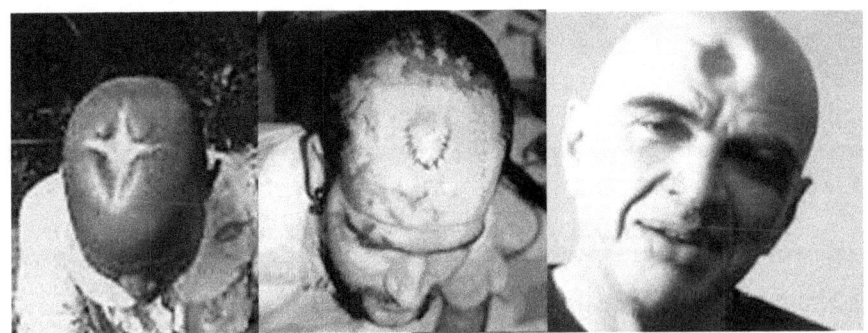

There is no way of knowing how many people lost their lives during trepanning experiments, but the practice continued because scientists told the population they would be better off with holes in their heads than demons or whatever they decided they could cure. Today, doctors certainly trepan individuals in preparation for brain surgery and for the relief of swelling due to brain trauma, while these methods are well proven by animal testing, clinical trials, and the like, a few early trepanationists simply decided to make the world better by relieving the brain of demons and similar evils without science. It probably started as a small group wondering how to make a name for themselves. They decided by "agreement" for fame, backing of the leaders of a country, national pride, and other key elements, the idea becomes more than a theory it becomes a law. One such action of this type of quasi science discovered the medical breakthrough called heroine.

Heroine Theory

To save the world, doctor's leeches were replaced with Opium, Cocaine [opium], and Laudanum [opium].

Leech medication, and other old remedies were modified by "agreement" science into simply making people not realize they were sick to cure them. I know some doctors still use the same science and hand out narcotics to keep their patients returning, but it is not real science. Heroine started out as the cure for all coughs and ended up as the cure for everything. Below are a tiny fraction of the varieties of opium curatives. "agreement" outweighed science and observation again.

Heroine Brain Salt would cure brain troubles, sleeplessness, nervous debility, excessive headache, and sea sickness. Cocaine toothache drops would cure toothache, Blood disorders, skin eruptions, and loss of energy safely. Heroine medication called tippicanoe used the popularity of the President to cure dyspepsia,constipation, stomach disorders, trepid feelings, feeble appetite issues, female complaints, malaria, and mal-assimilation of food,

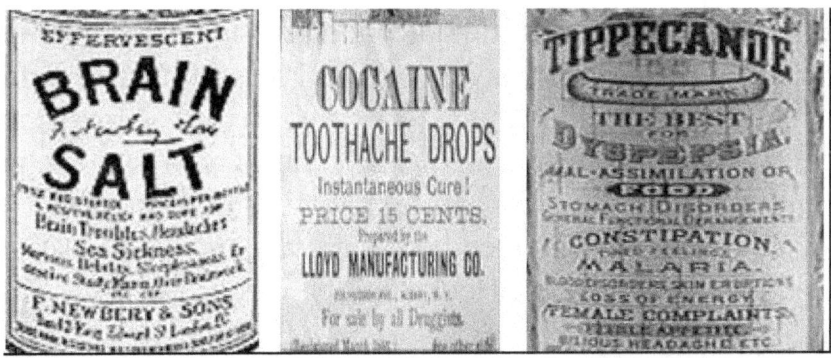

Heroine soothing syrup was marketed for children's teething. Heroine Vaporol was used as a vapor treatment for asthma and spasmodic affliction

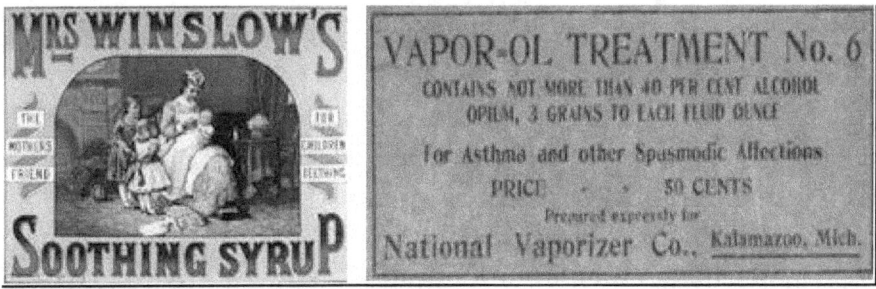

Heroine Cough syrups siad they would not affect other areas of the body and would make childrens's stomachs feel better in a harmless way. I wonder what sicentists believed harmess was.

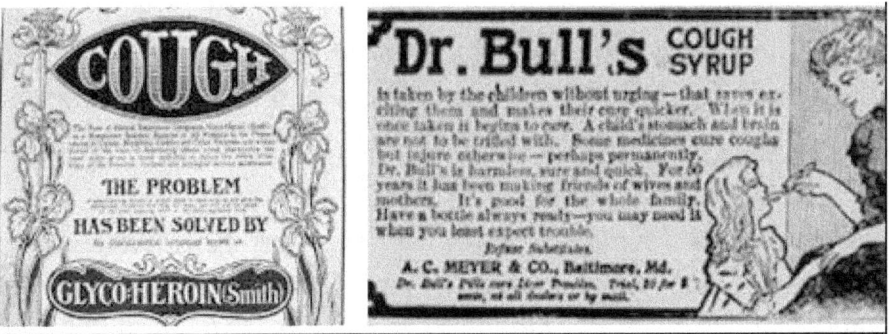

I can remember when I was given paragoric as a child and I had no idea they were filling me with heroine. It was called

the family babysitter as it made children very dossile. So that brings us to the first bayer asprin. Please notice the first Bayer asprin, middle image, was just heroin. Also shown next, heroine Daimiana medicine claimed invigoration and strengthening medication to eliminate signs of earlier in indiscretions. It invigorated both the brain and the nerves and cured impotency, and nervous debility.

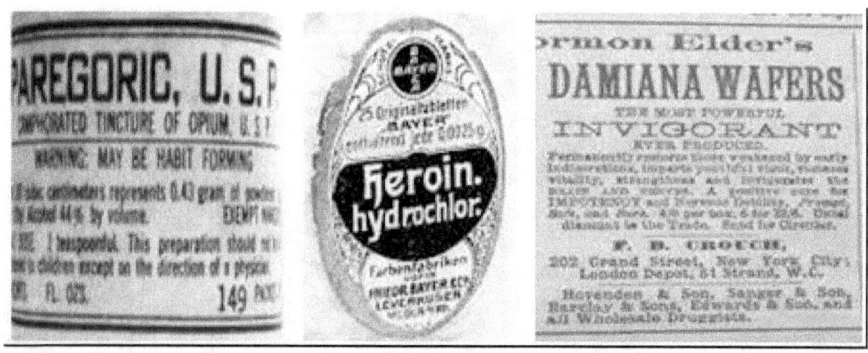

By "agreement" the addictive opium or heroine was widely distributed by barbers, confectioners, ironmongers, stationers, tobacconists, wine merchants, and doctors. It was easy to come by and many people took it, including numerous authors who became addicts such as Elizabeth Barrett-Browning, Lord Byron, Wilkie Collins, George Crabbe, Charles Dickens, John Keats, Percy Bysshe Shelley, and Walter Scott, Samuel Taylor Coleridge, Elizabeth Barrett Browning, and Charles Dickens. Most of them destroyed their lives just as many, many others had. Possibly the most famous of the tragedies or successes of heroine was Coleridge's partial poem *"Kubla Khan"*.

In Xanadu did Kubla Khan-A stately pleasure-dome decree: Where Alph, the sacred river, ran through caverns measureless to man- Down to a sunless sea. So twice five miles of fertile ground With walls and towers were girdled round: And there were gardens bright with sinuous rills, Where blossomed many an incense-bearing tree;

And here were forests ancient as the hills, Enfolding sunny spots of greenery.-------

His written "partial poem" obtained from an opium stupor is shown next. He would spend the remaining 30 or so years as an addict and even went to live with his drug supplying doctor for the last 18 years. To the right a special form of opium with Lead used to get an extra kick. If the opium didn't get you the lead poisoning would. It was very popular in Coleridge's England.

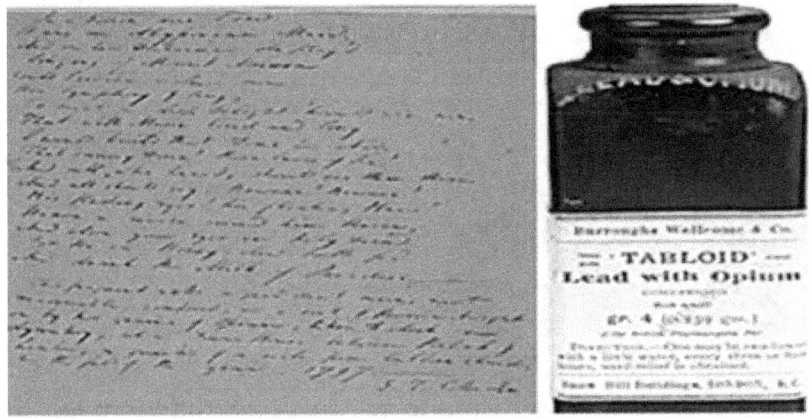

There was nothing heroine could not "cure". Soon just about everyone had heroine on hand at home to cure just about everything. Patients loved how well the medicine worked and came back over and over again to be cured. Today we have names like Oxycodone, Hydrocodone, and Fentanyl to hide the dangers of opioids and doctors are still treating patients by addiction rather than curation. The entire world looks better while medicated so it must be curing the entire world. As heroine lost favor, "agreement" scientists grabbed on to a new way to control the masses and give themselves power. This was done with Radium.

Radium Theory

Doctors and practitioners decided the entire world should grab hold of the miracle known as Radium. Home and doctor use of Radium curatives and treatments abound through the first half of the 20th century and beyond as scientists developed cures for anything heroine had problems with.

Radium was sprayed, smeared, injected, pushed down every orifice you could imagine as the "agreement" scientists told the world that they would essentially die without this miracle curative. I know you hear about things that are going to destroy the world by these same types of sinister people controlling entire populations by fear as they parade some TRUTH that their group made up to get them rich, famous, or even to promote some fake scientific agenda. Radium scientists, engineers and manufacturers had their day during the first half of the 20th century. Degnen's Radio-Active Solar Pad purchased by tens of thousands would eliminate sluggishness, high blood pressure, diseases of the stomach, heart, lungs, liver, kidney, constipation, gout, sciatica, and other similar ailments. For Men, there was the Testone Radium Energizer and Suspensory. [2nd below] "The Rectorotor" came along to add radium through the butt to cure constipation, prostatitis, piles, and similar discomforts. It was decided that Radium could cure just about anything. Radium suppositories would assure a better sex life.

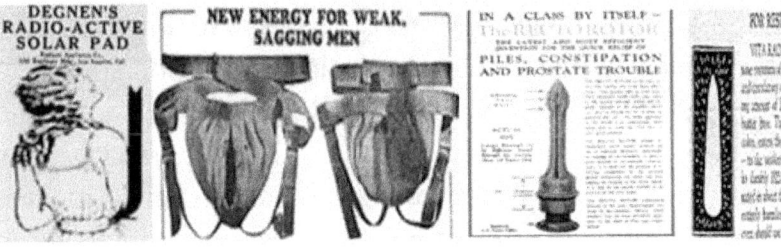

A tube was inserted to deliver radium cleansing for Bladder and Urethra issues as shown next left. [OUCH!] Other radium treatments had the hands and feet placed in a radioactive bath. Sometimes, just breathing radium in the air could fix your woes as shown next right.

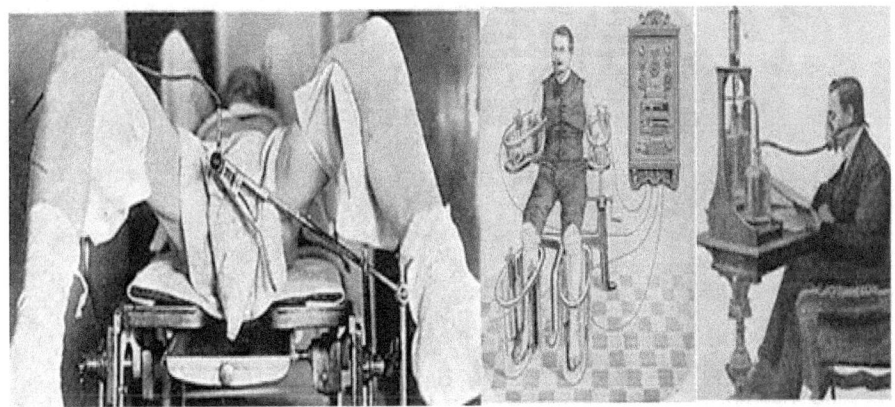

For **skin Cancer**, they came up with radium location modules for the skull or as a partial mask on the jaw [see last 2 images].

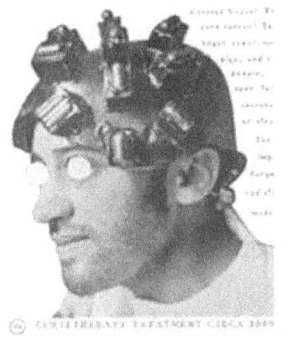 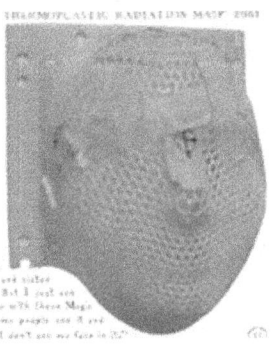 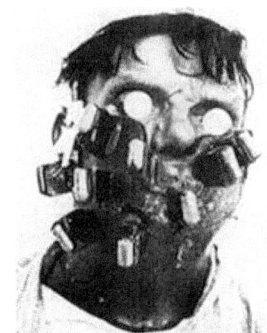

"Radithor" would **cure Arthritis**, Rheumatism, Neuritis, Gout and other joint issues. Radium slave could be spread to eliminate cancerous growths. Tho-Radia made a whole product line of perfumes, creams, facial powders, lipsticks and other beauty products that contained thorium chloride and radium while an interesting curative was the Radium chocolates by Burkbraun. Radio-Sulpho would dissolve all poisons from the body as a germicide, antiseptic, pore cleaner, and eliminated urinemia, rheumatism, swellings, blood poisons, and, of course, Cancers.

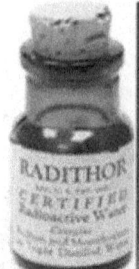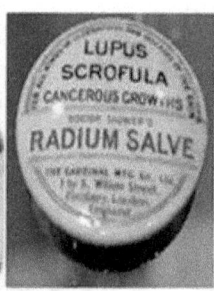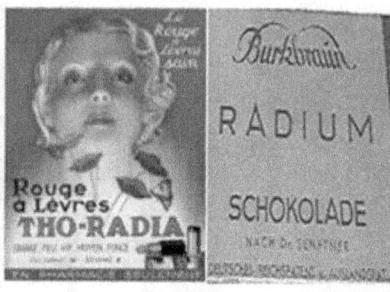

Radium Condoms by NUTEX assured a high percentage of accuracy. This medical device was patented in 1912 by R. W. Thomas and manufactured by the Radium Ore Revigator Co., which sold thousands of them in the 1920s and 1930s. It was basically a ceramic water crock lined with radioactive materials (uranium, vanadium, radon), lead and arsenic. Radium Hand Cleaner – "It Takes Off Everything but The Skin". La Parle **Obesity Soap** even eliminated fat with Radium. Drinking water was replaced with Radio-active radium infused drinking water to cure "Tabes Dorsalis", "Catarrh of the Antrum and Sinus", **Diabetes**, Glycosuria, Nephites, and other ailments in the hospital and home using the third device. If you didn't have a home fountain, you could use Radione tablets to add to water for energy. Radi-endocrinator papers were spread over the inflicted area to fix the endocrine glands.

Glen Springs Health Resort and Hotel opened up as a health resort with its highly radioactive Radium hot springs. Just relax and cure everything. At home, one would simply use Radio-active radium "XRay Soap" and radioactive Laradium **face cream made the complexion a thing of beauty** in just 20 minutes. Radium Hand cleaner was advertised as being able to remove everything but skin.

Radium tablets under the name brand Arium simply claimed to **make you a super-man**. Radioactive wool by Laine Oradium for self-heating clothing and Radium heating pads kept everyone warm and safe. Then there was "Caradium hair conditioner" that would eliminate grey.

For Eye problems, the "agreement" scientists came up with "Radio-bleu while Radio-X tablets could **clean the bile ducts**. Even teeth got brighter using Doramad radium toothpaste. For Smokers, we find radium infused "A-Batschari" Cigarettes to be lit by Radium "silent matches".

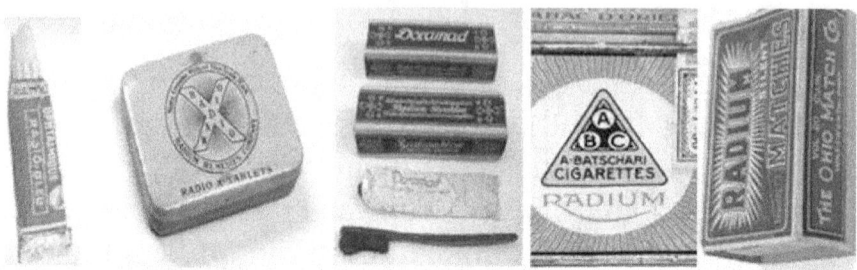

Multiuse "Radium Spray" could be used to polish your furniture and if a bug flies by you could kill it with the same spray. Almost the same Radium formula was marketed as leather dye and as metal polish. Radium "Gloss Starch" could even clean your clothes.

For Lunbago, Sprains, Bruises, Chest colds, swollen joints, and coughing, just take a little "Radium Radia". There was even reports of Radium **bringing back sight** and doctors would routinely use a radium compress or as an intravenous injection shown third and 4th.

The X-Radium foot warmer came along with a huge number of commercial aids including glow in the dark radium painted chandeliers, glowing jewelry, and glow in the dark wrist watches. I had one of those things when I was younger myself. Even glow in the dark Roulette became the rage.

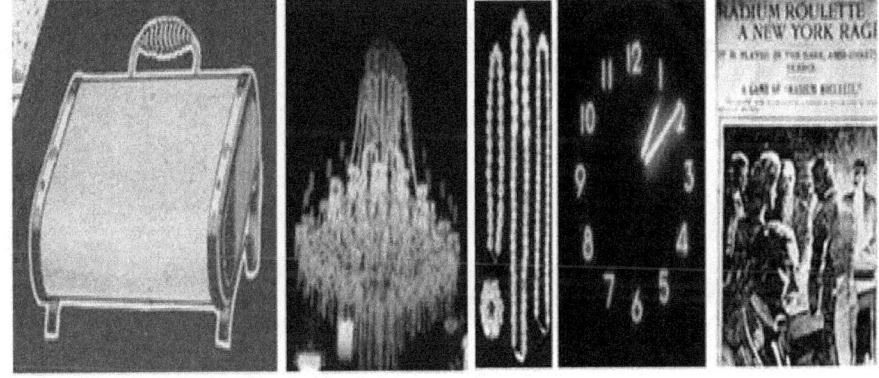

Soon radium death tolls began to climb. Those not dying immediately were sometimes disfigured horribly. I even got rid of my radium watch. You would think people have more sense than to drink something that was being sold as a bug killer, but "agreement" scientists can sway just about anyone and a new "agreement" science was just around the corner.

  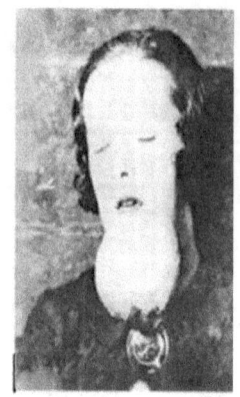

Science by "agreement" had claimed many more victims and all because we are too trusting when someone puts scientist before their name.

Just when you think scientists would not simply make things up to make them seem important, we hear about Evolution.

Anthropologic Misconceptions

Spontaneous Evolution Hypothesis

This crazy thing says that some hunks of amino acids were sitting on a rock one day and lightning hit the rock. Thousands and thousands of the same mutations occurred to change goo into life. From there, every time a cosmic ray hit a group, a whole bunch of animals would turn into a different type of animal.

Speaking of inappropriate testing and determination. Evolution is a neat way to show existence of animal life that causes no fear or thought by the "helpless" evaluator. The problem is that, most of the time; it doesn't work and doesn't fit the evidence. At best this unproven, untested determination should be called a hypothesis or a hunch, but it is taught in schools as almost a "Law". Let's look at some of the problems with evolution. Many call spontaneous evolution the "No-God Creation". After we look the "No God creation, we will examine problems with the "One-God Creation" theory. That's just as bad so don't think I'm trying to play favorites here. Any science can be twisted if you try too hard to make your theories work regardless of the evidence.

Too Many Animal Types

Too many animal types were spontaneously generated if an evolution theory was to be supported. We will see that after each destruction period, a huge influx of animals was apparently created <u>within an extremely short period</u> rather than the supposed "start over" that evolution theories would require. Each of the seven generally known major destructions recorded in history caused the extermination of well over 80% of all species of life and each time more animals emerged. Here is the strange part. The evolutionary life cycle didn't start over each time, but instead, many creatures spontaneously reappeared. To make things even stranger, a complete extinction record shows as many as 18 major extinction periods and many periods of minor extinctions during Earth's life cycle, which further exacerbates the problem.

Quick regeneration of life after extinction doesn't go along with evolution. Even cosmic rays have their limits.

Out of Place Objects

Many things are out of place in time. Man's footprints with dinosaurs, cups found in coal, walls found deep in the Earth, and many other artifacts clearly show that manlike animals were on the Earth many years before the time that evolution can support. A sampling of a few of these anomalies can be seen in later chapters. Don't be fooled into thinking that the artifacts are proof that the world is very young and testing methods are a sham. We will investigate some of the methods and reasons why a young Earth is not a better answer than the "evolution theory".

Out of time relics disprove evolution.

Missing Link

No crossover animals have been found. No matter how much searching, no one has found any evidence to suggest that a change in genus kind is possible- a horse has always been a horse and a man has always been a man. I know that many are saying that they have found plenty of crossover species like the partial bird partial reptile creatures [Archaeopteryx], but they truly do not represent a cross over. They really describe different separation of species. If we were looking for crossover people we might look at Charles Darwin himself as referenced to the Homo Habilis drawing to the right. Although they are similar, it does not mean Darwin is a missing link.

Natural Selection Model Anomaly

Natural Selection and survival-of-the-fittest models are sort of corollaries to the evolution theory, but they don't work either. Shell fossils show that during some ages they get big and then small and then big again at a later time period. Horse "evolution" also shows the same characteristic. The horse started off small, got medium sized, got smaller, and then larger. [weaker, stronger, weaker, stronger etc.] All other animal fossils suggest that the adaptation to the environment does not increase the capability of an organism to survive, nor does it make a superior organism, it just kills the organism off and miraculously a new type takes its place. Sometimes it's a better animal, but <u>many times it is a "less survivable" creature. By the way the timing method used to help make "uncontrolled Evolution" seem possible has been debunked for years now. What originally had been 65 million is not believed to have only been 120 thousand years.</u> The following chart uses the old nuclear decay methods. The 55 Million is really more like 110 thousand and the other dates have similar proportions except the last one.

Types of Horses and When They Were Here

Type	Time [yrs ago]	Size	Toes
Hyracotherium	55M	12 in.	Four
Orohippus	50M	14 in.	Four
Epihippus	47M	20 in.	Four
Mesohippus	40M	24 in	Three
Miohippus	35M	30 in.	Three
Archeohippus	24M	**18 in.**	Three
Merychippus	17M	40 in.	Three
Protohippus	16M	**28 in.**	Three
Hyracotherium	15M	40 in	Three
Pliohippus	14M	44 in.	One
Dinohippus	12M	**39 in.**	One
Astrohippus	10M	48 in.	One
Equus [modern]	4M	65 in.	One

Evolution Reversal Evidence

Why are no fishes now changing into amphibians, amphibians into reptiles, reptiles into birds and mammals, and monkeys into man? If growth, development, evolution, were the rule, there would be no lower order of animals. All animals have had sufficient time to develop into the highest orders. Many have remained the same; some have deteriorated, none have evolved.

If plants and animals all developed from a one-celled animal, such as the amoeba, why did the amoeba not develop?

Animal Variety Evidence

We are told that, excluding the insect and microscopic world, there are about 3,000,000 species of plants and animals today. About 1/3 of that number are animals and about 0.05% of the animals are going extinct every thousand years. That gives us about 1 million species of animals with over 500 species being lost every thousand years. Now let's assume that the last major extinction took place about 60 million years ago. From that comes a very serious question.

How many new species should have arisen in the last 5,000 years to support the undirected evolution theory?

If we start from 60 million years ago there would be 20 animal species doublings to approach the 1 million animal species we have today. If we assume that each doubling takes the same time period, then the last 500 thousand animal species would have sprung into life within the last 3 million years. If we further expand the 3 million years into 5 thousand-year blocks, we find that within the last 5 thousand years there should have been over 5 thousand new species being generated if we disregard the extinction of previously generated species. We would then have to add another 25 hundred new species to make up for those which periodically become extinct [the 0.05% per thousand-year figure]. <u>I'm sure everyone knows there have not been 7.5 thousand new species generated in the last 5 thousand years.</u> The number is very close to zero.

Darwin really wanted to believe his own hypothesis, but finally indicated the following: *"In spite of all the efforts of trained observers, not one change of species into another is on record."*

The Canadian Geologist, Sir William Dawson also wanted to believe, but soon, in his frustration, he said: *"No case is certainly known in human experience where any species of animal or plant has been so changed as to assume all the characteristics of a new species."*

Not having 7500 new species in the last 5 thousand years debunks the evolution misconception, and then there were people with dinosaurs.

Stone Bone Theory

> *How could someone find modern human skull bones that were fossilized if it takes hundreds of thousands of years to fossilize? The only reasonable explanation is the people with these bones must have been fossilized when they were alive as no one could have lived with the dinosaurs.*

Besides the examination of australopithecine bones and the like to determine men came from apes, other elements of testing should be accomplished, one is to examine hundreds of ancient texts including the Bible that tell us about *"the Giants of old"* [Genesis 6] and similar descriptions from the Sumerian and Babylonian, Incan [Peru], Mayan [Central America], Ughu Mongulala [Brazil] and others that tell us about the group sometimes called the Titans or Homo-Gigantus. Certainly, we can't get DNA from stone, but what we can recognize is that these bones, some of which are huge were on people hundreds of thousands of years ago. Petrified bone should be solid proof of the Titan humans that lived with the dinosaurs, but somehow, textbooks forgot the thousands of pieces of evidence, hundreds of footprints located on the same substrate as dinosaur footprints walking along prehistoric beaches, shoe prints, massive bones, manufactured goods inside geodes, objects found deep inside coal mines.

In 1981, Wilton M. Krogman, the internationally acclaimed bone expert identified the first "stone bone" as a human calvarium, a portion of a skull with the eye-sockets broken off. It should be noted that the cellular structure of bone is plainly visible in a Microscopic study of this specimen as well

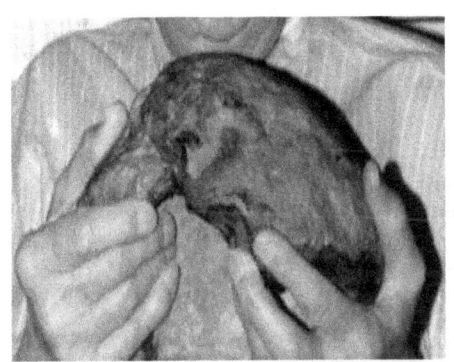
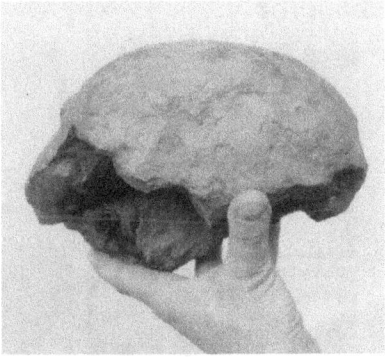

In 1982 a Researcher named Ed Conrad discovered in Pennsylvania petrified teeth inside the jaw-like area of solid rock. Then he found more and more specimens that bore the contour of human bone. You can just call it a stone head, but petrified skulls tell a tale. [See following image]

A CAT scan had been done of this particular specimen and revealed intriguing characteristics of a human skull jaw and joints. [See above right]. I know this makes you mad that people living with dinosaurs during the Cretaceous Era were ripped out of your history books and science information that would allow you to understand how humans got here better, but there are some who will protect their "theories at all cost even if they have to bury the past as we will see in the next anomaly.

More and more Modern Petrified human bones began turning up. The following collage shows a mall segment of this anomaly.

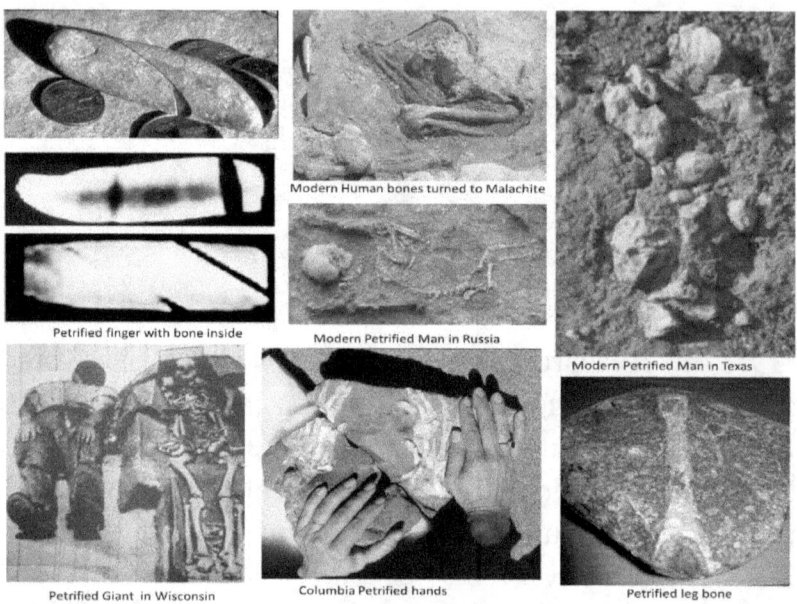

The first image shows a petrified human finger found in Wisconsin with a bone in the center as attested by x-ray. Then there was the petrified man found in Russia middle, followed by the Texas petrified man. Along the bottom row we find similar signed of extremely ancient humans living in Wisconsin and in Columbia South America. Please notice that ancient people from Wisconsin were huge. Many modern skeletal elements like skulls, leg bones, teeth and other elements crumble into dust when exposed to the atmosphere, they are so old. But some remained showing Cro-Magnon human skulls that had petrified teeth as shown next.

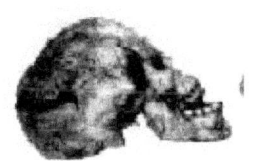 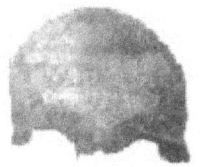 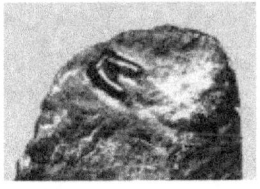

With all this evidence and much more, modern scientist say Evolution controlled advancement of humans so there were no humans during the Cretaceous time. Therefore; these are not bones.---- but what about all the other stuff?

Advanced Mesozoic People

> *Unfortunately, there was another little problem with evolution. Modern looking human artifacts have been showing up from as far back as the middle of the Mesozoic Era. Put that in your evolutionary pipe and smoke it!!*

We have found hundreds of items. Even if some are mistakenly identified; the list keeps getting longer.

Ancient artifacts are encased in multimillion year old coal, rock, or even geodes like silver chains, bells, cooking utensils, thimbles, hammers, toys, artwork, and even electrical devices. A small sampling is shown below. These show a high level of sophistication of these Mesozoic people

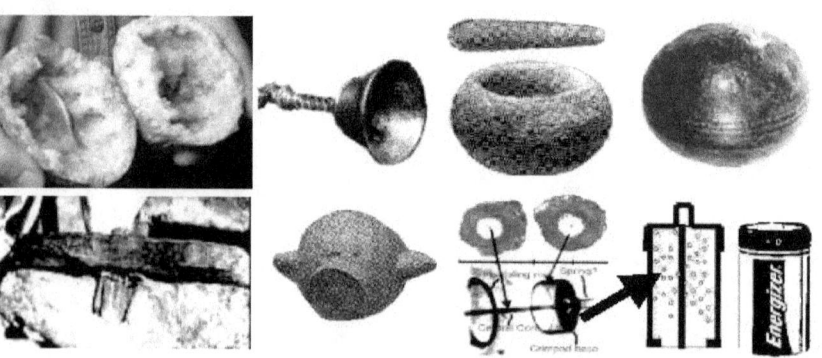

Footprints and shoeprints are found with dinosaur prints. One famous shoe smashed 2 trilobites as the person wearing them walked. [Trilobites have been extinct for over 100 million years.] Around the world in Russia, Texas, Kentucky, Australia, and other places, we now know that dinosaurs and humans walked on the same beaches enjoying the same view.

 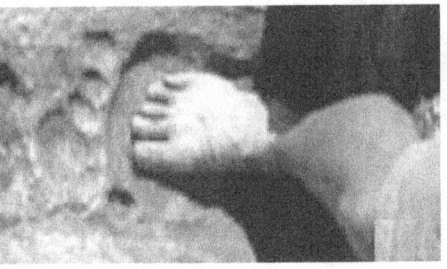 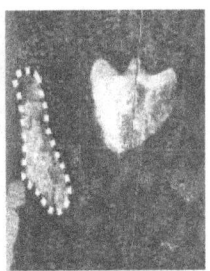

We have found that these people not only wore shoes, but also, they had clothing that was sewn together rather than having skins draped on their butts. The evidence of the shoe that crushed trilobites is shown in the following image.

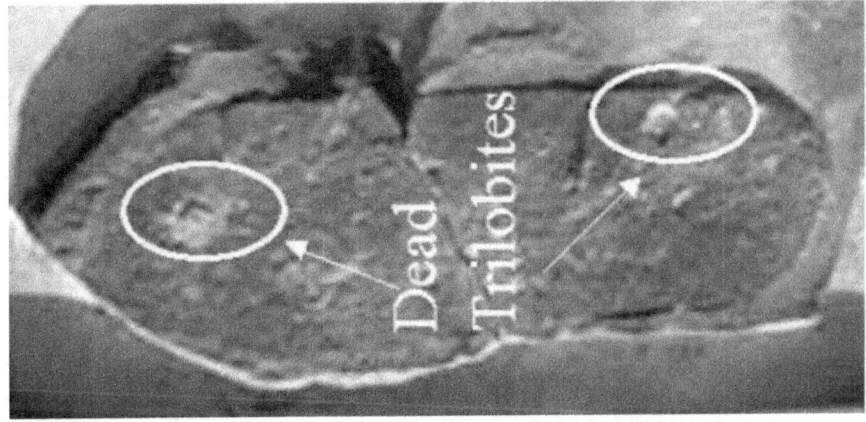

On an on we can go to tell us that human evolution was not what we have been told. I would like to get into a more reasonable explanation, but I must continue with the stupidity scientific theories or this book will never be finished. I will, at least present an overview for you to consider.

While this is at odds with spontaneous evolution, there is a problem as many of the animals are not designed well and God wouldn't do all this experimenting. Something else is, most likely, the answer.

Experimentation Theory

This theory is simple God would not make the mistakes found in the Animal world. Someone else must have designed MOST of the animals. What we have is evidence of experimentation.

Twisted Nose Continues our Anomalies

Sometimes "evolved features" are recognizable as mistakes. A Dinosaur mistake to be considered would be the long-nosed Dinosaur, with a trombone in his head. As shown in the following picture his nose is over four times as long as his head and was curled back on itself. It was completely useless and there is no evidence to suggest that other dino-features evolved from this mistake. The nose couldn't be used as a battering ram like the horn extensions of other animals. It was just a long nose. For those who would suggest that this is the father of the elephant with his useful nose/hand, it would be improbable that the thing would have evolved from this characteristic to the dangly one of today. This type of mistake probably didn't come from an omnipotent creator, nor did it come from an evolutionary process. We don't even think it could make music. We will talk about a third option.

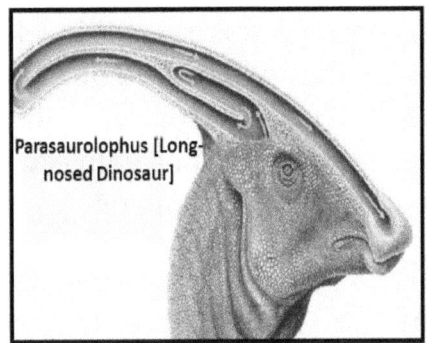
Parasaurolophus [Long-nosed Dinosaur]

Human-Parasaurolophus

If we assume that God would not need to experiment, and that the natural selection concept doesn't support reduced capability outcomes, we must strongly consider ancient human manipulation during an ancient time. That idea is unsettling and confusing so people keep on using evolution as a way to not think about a reasonable truth.

Scientists have looked around and found all sorts of strange animals and tried to piece together a crazy story that would force fit all the mistakes that were made during the genetic experiments. While our schools instruct us like an spontaneous and uncontrolled evolution could work, it has never been able to explain any of it. It can't explain why ocean clams are less evolved today that during ancient times or why the smaller animals survived while large animals could not. It cannot explain why amoebas never evolved while man's brain expanded almost 1000 percent over a mere 100 thousand years. It can't explain why rudimentary tools were used exclusively up until about 40 thousand years ago then, all of a sudden, the ancient tools completely vanished around the world and complex tool took over in a matter of years. In fact, nothing can be explained by the concept.

Possible Evolution

For some reason, there are still people who say that there are many things that show uncontrolled evolution and survival of the fittest controls history. Certainly, there is a level of change inside a particular species, but when it comes down to it, cross species changes and even interspecies oddities cannot be answered by evolution alone. I'm not going to get into many of these, but a few are probably warranted.

Stupid Bird Heads

If evolution and survival of the fittest were the only controlling factors for survival and existence, why are there so many surviving animal types and why are there so many traits that are generally so useless. By the way, why would animals be so vastly varied and strange unless a group was "experimenting" with different things? Let's look at birds over the years. All those topknots and things were not helpful to those birds. They were mistakes and the creator God would not have made mistakes. Take, for instance, the Toucan. What in the world is that beak useful for? It makes it more difficult for it to fly and survive, but there it is- a big chunk of useless beak. Anyone who has ever been bitten by a toucan knows that the bird is almost totally defenseless when considering his "beak power". What a wimp! Below are some of the reconstructions of bird ancestors. Can you tell which came first? Most of the unusual characteristics are completely useless, but somehow the "bird" survived during the preflood days. I like the one in the bottom middle that big bone topknot makes it look like the woody-woodpecker cartoon character and I guess it could have been used as a rudder, but if he turned his head to look at something while flying it could be very dangerous. I suppose one could say this bird survived because the topknot was used for a special purpose, but most should recognize that the topknot was a mistake.

Seriously, these heads show major mistakes in genetics that survived just like the Toucan as "survival of the LEAST fit" shows mistake after mistake.

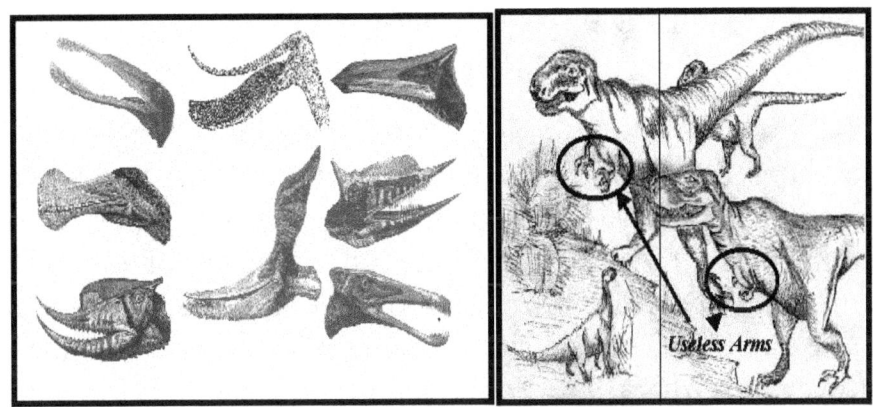

Stupid Arms

Speaking of mistakes with arms, we have got to look to the Tyrannosaurus Rex as shown previous right. In the recent past, this animal was determined to be the highest form of predator, but now we know differently. His arms were too short. Although he could run, if he fell it was an almost impossible situation for him. While it was possible for him to get back up, the energy expended would have been tremendous and many of these short-armed mistakes would not have survived. The new picture of the Rex is that of a clean-up animal, like a vulture. He was big, strong, and fast, and he had tiny little arms. I saw a documentary indicating that the tiny arms reduced the weight on the legs so the tiny limbs were reasonable and necessary. Anyone looking at the huge, heavy, hulking head and saying the arms got shorter to allow the being to stand is somehow not seeing the big picture. There can be no mistake. T-Rex was another thoughtless genetic manipulation experiment. This time the geneticists were in North America.

Stupid Legs

The dinosaurs and birds were weird, all right, but what about other animals of today? Let's take grasshoppers and butterflies, for instance. Their legs are being used for just about everything and it doesn't seem right.

Stupid Leg Tasting

Who would create an animal with "taste buds" on its legs? Now it seems dumb, but when the butterfly was created, experimentation was done on every level. It reminds me of the haphazard DNA experimentation we have done today. I remember one experiment introduced in Biology class. In that experiment, the scientists forced wings to grow in place of eyes in a fruit fly and a human ear to glow on a poor rat, as shown below.

I'm pretty sure that some future evolutionist will come up with a reason why the wings were more important than eyes and the ear helped the rat hear. Whatever they do it will be to show the mutation was an evolutionary requirement. The truth, however, is that today's geneticists are developing animals without regard for the outcome just like they were doing during ancient times.

Stupid Leg Hearing

Years after the T-Rex experiment, "creators", or should I say experimenters, got to the grasshopper and decided to put its *ears on its legs*. Not only does that seem like a crazy place to put them, just think how loud it sounds to the grasshopper when he makes his leg scraping sound. The grasshopper certainly must hate this mistake. I know you can probably come up with hundreds of oddities that make no sense with respect to any type of evolution process, but these are funny and appear to be jokes played on the unsuspecting animal. If you really want to see a weird one, just look at the squid.

Stupid Squid Mistake

You can't talk about mistakes without bringing up the giant squid. Portions of these animals have been found which indicate sizes of over 66 feet long and 40 tons are common with the odd 18-inch diameter sucker marks found on whales strongly indicate that animals larger than 100 feet are in existence. Even after million years of "evolution" they generally are characterized by the following components.

- No bones
- Their blood is the wrong color. Its blue instead of red.
- They are fast enough to stay away from whales, but so stupid they simply try to hide in the cave-like mouth of their enemy. They hide almost all the time, but when you hide in a whale's mouth, you can expect the worst.
- When not hiding, they swish themselves through the water backwards which makes sure that the huge eyeballs can't see anything.
- They may even produce their own light to add to their strangeness.
- To top all this off, they are one of the most primitive and oddball beings with respect to mating.

Hectocotylus Mistake

A squid's penis swims so scientists named the thing hectocotylus. At first this swimming capability seems like a really neat way to insure procreation, but once the penis is gone what is the boneless creature supposed to do? The squids have a huge penis, at least the males do. It's up to 4-foot-long, but there are two serious problems. Many male squid seem to have no sexual organs at all [I bet you can figure out what happened.] and the female has no sexual opening. It is believed that squid sperm is injected directly through the skin of a female and that is where the hectocotylus comes in.

These hectocotylus things were first thought to be gigantic worms of the sea, but now we know that some species of squid, not only inject sperm, but the entire penis shoots out of the squid and swims toward it new home so that the female can take it and have a substantial cash of sperm for later procreation.

One squid's loss is another one's gain, or should it be finders keepers losers weepers? Possibly, it could just be, "Swim to the female you stupid penis."

No matter what you want to believe, this penis swimming thing is not because of evolution. It is because of a jokester from our ancient days.

Please remember that the Bible calls the squid one of the Abominable or unclean animals of the earth.

The Sad Whale

One of the mis-creations, we are told in our Bible was the Whale. With the largest brain of all animals, a mammal, and having a massive body. The inventors of this abomination set it out to live a life of horror and misery. While they had to breathe air, the genetics engineers, want to see if adding human brain growth and understanding into sea dwellers would make a smarter fish. Instead, a whale must stay in the sea so it won't crush itself, has no arms or legs so it can only swim, and to make it more, a baleen was made inside its mouth so it would have to eat millions and millions of tiny krill shrimp just to survive.

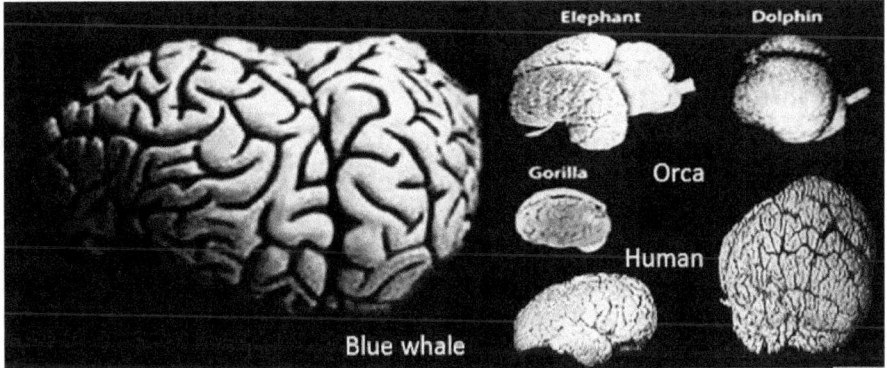

All day long the whale moans in agony and we interpret it as singing. We can imagine the inventor got the equivalent of the Nobel Peace prize for his great accomplishment. The previous collage shows relative sizes of brains. At first glance, the dolphin and Whale have more brain matter with the whale having many times the intelligence of humans and there is nothing he can do about it.

The Sad Dinosaurs

We are now finding Pleistocene scientists not only manipulated DNA to bastardize animals, they also re-created animals that had become extinct. Dinosaurs were remade during the Pleistocene.

We know this for 2 reasons.

1. Unfossilized remains of many dinosaur types showing they lived less than 30 thousand years ago have been found.

2. Some of these dinosaurs are very radioactive showing they lived during the Young Dryas when the entire Earth seems to have been radioactive from war. [We will look at this later]

Unfortunately for massive dinosaurs, the world was different than during the Mesozoic. Our planet no longer was spinning as fast so huge animals like dinosaurs would have been much heavier than they were during the Mesozoic. This means they were almost as miserable as the Whales. Every year we find more and more unfossilized remains of these unfortunate dinosaurs. One seems to have been a favorite was the T-Rex.

Triceratops, Hadrosaur, and T-Rex-Triceratops and Hadrosaur bones from Montana were tested for Carbon 14 two different dating labs both said that a triceratops registered an average of 31,000 years old and 23,000 years was the date for a Hadrosaur. Some researchers cut open a number of T-Rex bones and found masses of soft material that had not fossilized yet. Meaning the animal was less than 50,000 years old. The scientific community went berserk. This was a lie; this was a mistake; this was an anomaly. Soon all the back pointing was of no use as more and more finds showed the same thing. Some T-Rexes lived well past the time of the reported extinction. An example of the elastic material is shown next.

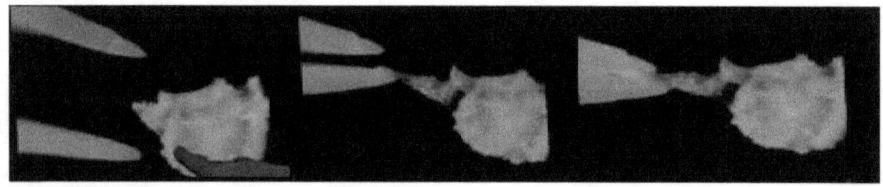

Duck-bill Dinosaurs with soft tissue- Some flexible portions from a leg bone are shown below right to increase the Pleistocene dinosaur count.

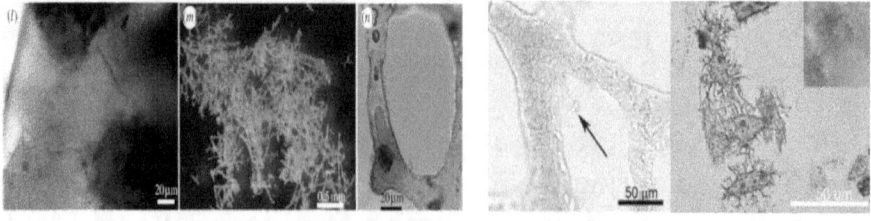

Many of the soft tissue dinosaurs being found are radio-active as well. T- Rex skeletons, many times are painted with heavy lead paint so people can view them safely. Interestingly, the 16-pre-Pleistocene dated nuclear reactors found in Gabon, Africa is missing enough fusionable material to have powered the entire city of New York for a year. Some suggest this material loss has something to do with a massive spike in world radioactive fallout during what is called the Young Dryas, 11 thousand years ago.

We can assume War just before the end of the Pleistocene as the Earth axis shifted and almost all life died 10 thousand years ago. That being said, pockets of humanity survived and so did the biologists.

Continued Bio-Experiments

"To help humanity", Holocene Biologists remade dinosaurs again, but this time smaller ones. A question might be can we modify the biology of animals today to help mankind?

Let's start this section with a huge amount of pictorial evidence of dinosaurs around the world.

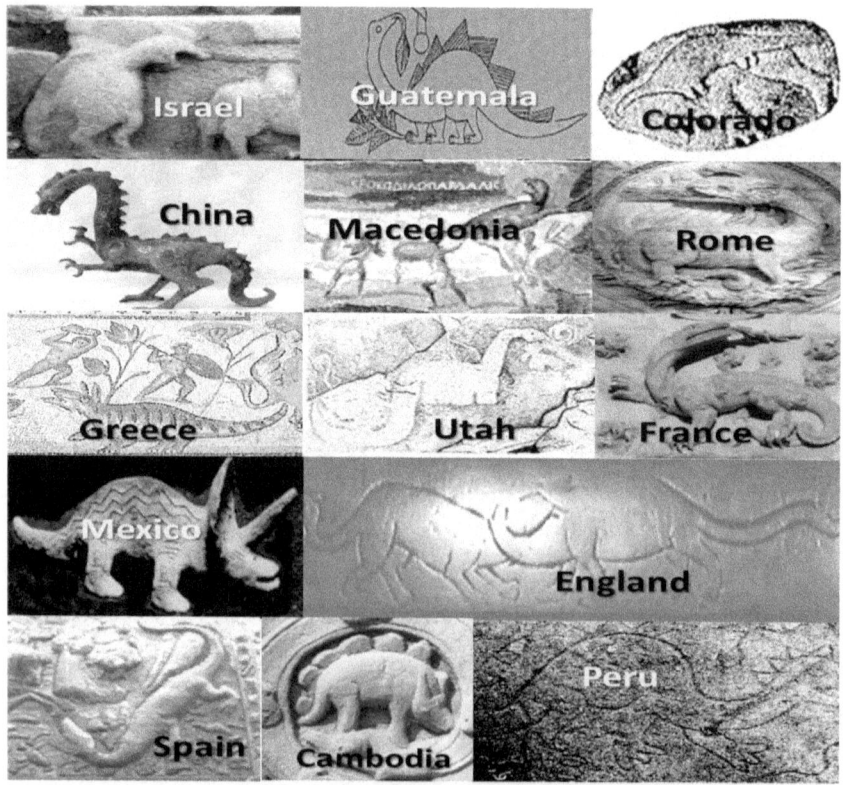

Some tell us there are still some of these monsters around and it doesn't matter much. The last chapter of Daniel describes how Daniel was asked to worship one of these monsters. Instead, Daniel made up a special concoction that the "dragon" ate and it, essentially, exploded. This made the priest so made they convinced the king to put Daniel in the Lion's den. In both Sumeria and Egypt we find that they had pet dinosaurs shown below around 5 thousand years ago. Egyptian dinosaurs are to the left, Sumerian ones are shown right.

It is believed that the scientist that had created these "Mis-creations were killed in a massive worldwide war called the Bharata War that ended 3100BC, but are these same types of mis-creations still being continued in the false name of science?

Mis-creations of the 20th Century

If you thought the above was getting so dangerous that someone should stop the genetic carnage that is happening every day and could soon spell disaster, let me review for you something that is even more scary as these same "scientists", and I use the name loosely, now making half human half animals all in the name of science and protecting people. If they make these half humans, they can get the organs and other things by simply not considering the new "thing" human and killing him. To the left in the next graphic is a picture of a rat growing a human ear on its back so don't go thinking that the Greeks were all wet in their descriptions of half-animal half-man beings. I read somewhere that the latest designs include a 15% human thing. The mouse on the right is being controlled by command pushed directly into his brain.

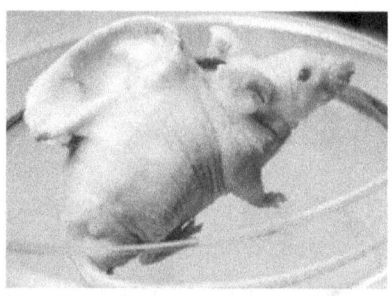 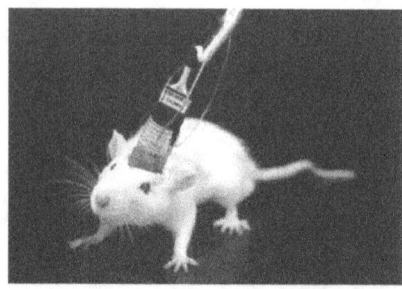

Maybe we should not get too worried, but remember, according to ancient texts around the world, God aided in the destruction of almost everyone during the Pleistocene Wars and again he aided in destruction during the Bharata Wars 5500 years ago. Each time well of 1/3 of the population of the world died if we believe historical writings of these events. OK! God may not have killed them directly, but they died just the same. One of the main reasons during both wars was that <u>each group started modifying animals like we are doing today.</u> The scientists during the Pleistocene Age, evidently, were the worst at not considering the negative side of their experiments and by the end of that age almost all the animals were considered abominations to God. The Bible even lists the experimental creations as "unclean abominations". The Zoroastrian historical references claimed the creations were the mis-creations of Satan. The book of Enoch also describes how the entire group that was "more" responsible for these "miscreations" was killed before the Pleistocene Extinction and worldwide flood, 10 thousand years ago and we are not listening.

Miscreated Human Hearts--Italian scientists have modified DNA to make a pig grow a human heart. The partial human has since been killed.

Miscreated Human Kidneys-If pigs can do it, so can mice. The Israeli's copied the Italians and made a mouse grow a miniature human kidney by introducing HUMAN stem cells. The kidney produced urine and the whole bit. It was just tiny. You are right believing that the scientists killed the partially human subject without remorse to get the kidney.

Miscreated Downs Syndrome-We have now produced a mouse with human "downs syndrome". You guessed it. Human DNA was used.

Miscreated human Nerves-Researchers in San Diego have designed mice containing fully functional human nerve cells as a novel way to study and potentially treat neurodegenerative diseases such as Parkinson's and Alzheimer's. The study is not necessarily good for the partially human mice.

Fun with fruitfly DNA- Below are some of the hundreds of mutated fruitflies being made as fast as a fly can mate. The first has an extra set of wings, then no wings, an eye on his leg, then useless wings followed by legs on his face instead of antenna, and then the no eye fly.

Human-Mouse Hybrid-Don't think these experiments are not known by the government either. On February 13th, 2005 the US Patent Office rejected a request by Stuart Newman and Jeremy Rifkin for a patent on a human-mouse hybrid.

Stanford's ethicists have tentatively endorsed the idea to endow mice with "some aspects of human consciousness or some human cognitive abilities." The British academy and the U.S. National Academy of Sciences have likewise refused to permanently restrict the humanization of animals.

Kill to get an Ear-There is the genetically engineered ear on a mouse as pictured previously, the mouse looks less than comfortable and I don't know what it could hear. I am pretty sure of one thing. This quasi-man will have to be killed to get the ear.

Miscreated Mental Retardation-Adding in human DNA into a mouse and modifying the "retarded" component is now being done to hopefully eliminate mental retardation. The mouse shown has been "de-retarded". Please ignore the huge cables coming out of his head.

Miscreated human blood-I haven't even begun to tell you about the dozens of Half plant- half human things that are being generated all to HELP humanity. One such half plant half human thing is a manufactured tobacco that produces human blood components. [I guess they get you to smoke to get a transfusion!!] To get the blood stuff they have to kill the defenseless quasi-tobacco guy.

Remote Control Rodents-As I showed before, electrodes implanted into the brain and a remote control, force this rat to turn left, right, climb a tree, jump, and stop on command. Just think what the scientific community has in store for people. [See next left]

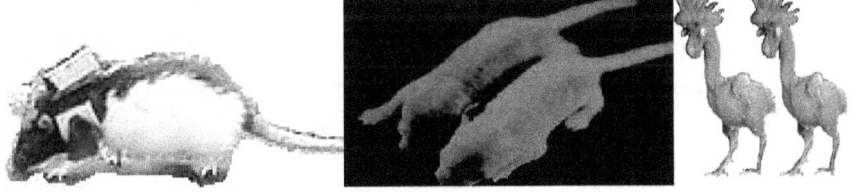

Miscreated hair-They altered hair to make it glow different colors. [The glowing puppies in the preceding graphic are only one of the many glowing animals that have been created.]

Genetically Bred Nakedness-We are not talking about people here, but instead we get naked chickens. These genetically engineered chickens grow faster in hot climates, contain less subcutaneous fat to hold the feathers, require less ventilation in the summer, and are cheaper to produce because they don't require plucking. [Oh yes did I mention they're ugly? See previous right.]

Miscreated Huge Brains-See if this doesn't sound familiar. Researchers in the UK grabbed some rats to see what they could do. Now they have modified a rat to grow a larger brain. Anyone who has seen the Pinky and the Brain cartoons knows what will happen with these guys, but now it's not a cartoon.

Glowing Rat Hair-In the United States, scientists said, "We can do something with those rats!" They mixed tropical fish and jelly fish DNA into mice DNA to make mice hair glow green, red, or blue. That was pretty good, but one scientist said, "Let's turn them into robots."

Mechanical Brain Transplant-We don't have brain transplants YET, but in March 2003 prototypes of the world's first brain prosthesis; an artificial hippocampus, was revealed to the public. By using this electronic device, scientists believe that defects that come from stroke, epilepsy or Alzheimer's disease may be treatable. The device simply reads information the brain normally sends to the hippocampus and it then sends out signals similar to those generated in the hippocampus. It's sort of a drop-in replacement. It will not only affect memory but also a person's mood, awareness and consciousness. It will make a "new person" and how much difference there will be is not known. The bioengineering cyberneticists hope it will be a better person and I do to. That hippocampus is a strange word isn't it? I'm glad God didn't make a mistake and put a hippopotamus in our brain. I also hope the geneticists don't make that mistake.

Miscreated Tooth Decay Bacteria- They manufactured tooth decay bacteria that won't cause tooth decay. [Now those are lazy tooth decay bacteria.]

Miscreated genes- They manufactured gene that eliminates muscular dystrophy in mice.

Miscreated Cardiomyopathy-They created cardiomyopathic mice so they could try to cure this human only disease. [I'm sure the mice appreciated this one.]

Miscreated human skin- They manufactured skin that was transplanted onto 45% of a baby's bad skin.

Miscreated increased life-They manufactured worms that live 50% longer than others.

Miscreated Sight- They genetically altered blind dog that could see again.

Miscreated eyes- They genetically altered sightless fish that now have eyes

Miscreated bones- They produced bone from skin and gum tissue

Miscreated fat burner- They developed an artificial drug that burns fat faster in mice.

Miscreated glowing skin- They genetically altered rabbit that glows under a black-light.

Miscreated ears and eyes- They genetically grew frog eyes and frog ears without a frog

Miscreated venom- They have genetically grown snake venom cells without a snake.

Miscreated pig-fish- They genetically altered pigs that now provide Omega3. [No more fish eating required.]

Miscreated fast dying fish- The altered a fish that grows faster, mates more often BUT his offspring dies more quickly. [This shows genetically manufactured animals could accidentally wipe out species.]

We can assume where there are a few successes, there are many mistakes. This list does show that cloned and manufactured animals and control of genetics are both becoming common place just like before. We can imagine the monkey-toads, goat-men, walking snakes, human faced monkeys and all the rest, but scientists keep mistakes in a back room somewhere until it is too late. With that, let's back up to the Tertiary period as genetics engineers make a new human we call Neanderthal.

Neanderthal Theory

I'm sure you know this one from the Geico Commercials. It states that from about half a million years ago until 40 thousand years ago a group of subhuman people lived and grunted themselves into oblivion. As they were part ape, they did not act human and their massive brain-was never used.

Neanderthal skull and reconstructed face is shown below. From his DNA, we believe a Neanderthal's hair was reddish and he was most likely freckled. He had a tenor voice, certainly used some type of language, and his brain was over 10% larger than ours.

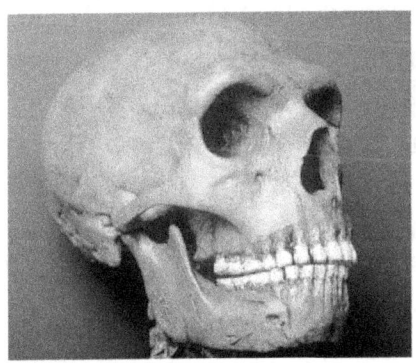

Before we look at the lunacy of the grunting Neanderthal, we need to look first at the Homo-Erectus. Some call this "reinvented" human the 6th day man as referenced in "Genesis" of the Bible. Although, at a quick glance, Neanderthal looked very similar to the Homo Erectus, his brain had jumped in size by over 50% over the Homo

Erectus who had just had his brain jump in size by a huge amount over the human-like ape we call Homo-Habilis. I'm not getting into the differences between true humans and the partial ones here, but just imaging the brain size expanding like blowing up a balloon. Neanderthal now had a brain larger than humans of today. Oops! That sort of messes up survival of the fittest, humans get smarter through evolution concept.

Neanderthal DNA

The search for viable Neanderthal DNA has come to some success, but the findings were not what the "normal scientists" wanted to find. Researchers have concluded that DNA strains of Neanderthal and modern humans were significantly different. They go so far as saying that Neanderthal had "ALIEN GENES" in their DNA meaning modern humans DID NOT have Neanderthal as an ancestor. It was also noted that Neanderthal had more brain capacity than modern humans. An additional item that has not been found, but is expected is the Neanderthal Baculum [I get into this in just a bit but we can simply call it a penis bone to foster familiarity right now.] Think about it the human that was supposedly our closest relatives were not even close to us. Smarter, alien in nature and possibly carrying a penis bone, this type of human changed overnight to become what we call Cro-Magnon. I'm not going to get into the Biblical significance of this transformation, but I will tell you that modern concepts of evolution cannot provide for this transformation. Many other anomalous features make scientists wonder about this strange human's appearance, but I think there is a logical answer to the jump in what some would call evolution. The Neanderthal was a hybrid offspring of the union between 6th day man and the ancient people. I know that sounds like the easy "cop out" answer, but there are many bits of information to lead us to this conclusion and none of them involve Neanderthal's grunting. As shown in the table following, there were marked changes that spontaneously occurred in this "new" human. The changes don't make sense. That is if you don't consider the introduction of crossbreeding with the modern humans of the REALLY ancient times.

Characteristic found in Neanderthal	Erectus
Found with "Alien" genes that didn't come from Apes	no
Brain size jumped by 50 percent almost overnight to 1200cc	Only 800cc
Got an elongated Brain [for increased motor skills]	Round brain
Began to live in villages	no
Began to protect the sick	no
Began to bury the dead	no
Began to have religion	no
Had an alien gene not found in modern humans	unknown
Had Big Pronounced Nose	no

OK! The big nose doesn't show advancement, but I would imagine that the Neanderthal were proud of their protrusion and ostracized the Homo Erectus for having the apelike flat nose. The petite nose showed inferiority and they wore it with pride.

Neanderthal's Alien Gene

If you noticed there was an alien gene or 2 in the DNA. This is really strange and needs to be explained as we have found the same thing in the Homo-Capensis [long headed human] skulls found in Paracas Peru. Over the last 150 years, scientists have struggled to unravel the mystery of the

Neanderthals. Of course, their conclusions are very limited because they tried to separate science and religion. The first significant discovery was made in August 1856. A partial skeleton was found at the Feldhofer Cave in the Neander Valley, in Germany. This was the find that gave the species its name. Since then over 500 individuals have been found from over 80 sites in Europe, the Middle East and parts of Western Asia along with several hundred thousand stone tools. It wasn't until 1997 that a small scrap of DNA was discovered which showed "Alien Genetics" with respect to our own DNA or that of an ape. Neanderthal could not have come from either group "exclusively" as neither group contained the portions of DNA found in Neanderthal. This unexplainable feature greatly confused the already misaligned groups of Paleoanthropologists. Besides the physical elements and no common ancestor, here are some of the things we have been told about the Neanderthal species as a social animal with a strangeness.

- Some of them were cannibalistic as determined from a Moula-Guercy, France site evidence.

- There was hybridization between Cro-Magnon and Neanderthal about 25 thousand years ago as evidenced from a recent Portuguese find showing many ambiguous characteristics.

- Even though the Neanderthal, as a species, lived for many thousands of years, they continued to use rudimentary tools and weapons.

- The majority of the "pure Neanderthal" stayed in the European area.

- They began the custom of elongating their heads artificially, showing a strong reverence to some type of longheaded humans that came before them.
- Long headed [homo-Capensis] human skull found in Paracas, Peru were determined to have "alien genes" making them similar to Neanderthal.
- Homo-Capensis humans are believed to have survived the Pleistocene Age and the long headed, giant features have shown up in Russia North and south America, Middle East, Egypt, and Europe. It should be noted the Paracas group were redheaded.
- Rather than in Europe, similar DNA characteristics with Neanderthal are found more in the Americas than in Europe.
- Neanderthal DNA seems to be similar to Denisovan DNA but Denisovan DNA characteristics are almost exclusively found in the area associated with Melanesia and Australia.
- DNA studies have concluded that ancestors of Neanderthal did not come out of Africa, but some may have gone to Africa 20 thousand years ago.

Neanderthal Brain

While we are on the subject of Neanderthal heads, we need to discuss brains, because, like I said before, Neanderthal had a larger brain and the brain was shaped substantially different than the "Modern Man Brain". The Neanderthal brain was much longer as shown by the elongated skull on the right as compared to the modern skull on the extreme right.

Modern   European   Neanderthal

The chart following shows the general progression of brain size to humanoid type and mean period of existence. The vertical lines represent the range of brain sizes determined for each of the subgroups. Note the sharp slope as brain sizes began expanding faster and faster until the Neanderthal.

By the way, this chart does not include the anomalous dolichocephalic giants [longheaded people that were mimicked by the Neanderthal.] Certainly, all the theories you have heard about until now have told about the longheaded people. The Hebrew word for "longhead" was ANAK, but scientist call them Homo-Capensis. King David killed the last remaining Anak that he knew about during his reign 3 thousand years ago and the Adena killed the last of the Homo-Capensis in North America about that same time, but that is another story.

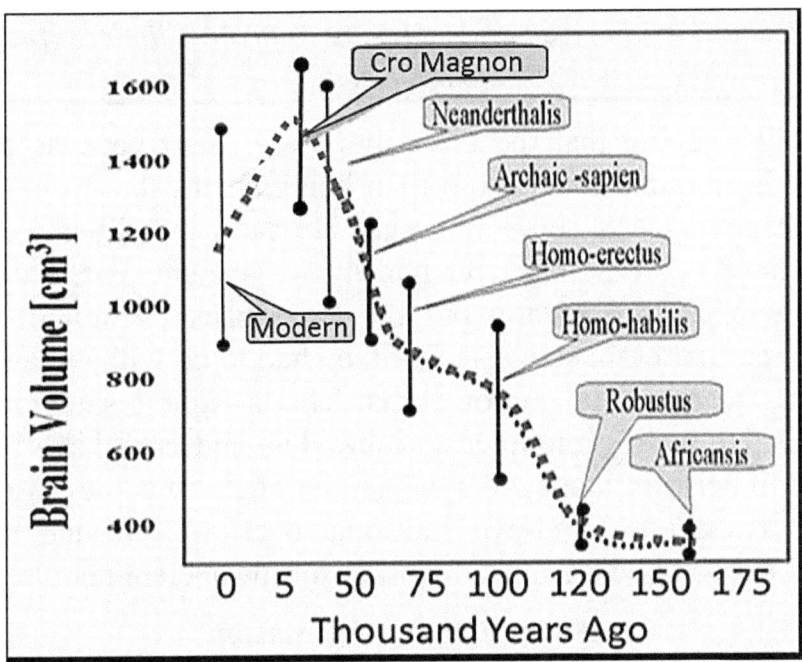

This whole" long flat brain" and the "round brain" anomaly should be investigated further. By the way, a group of the "Longhead skulls" found in South America are shown below.

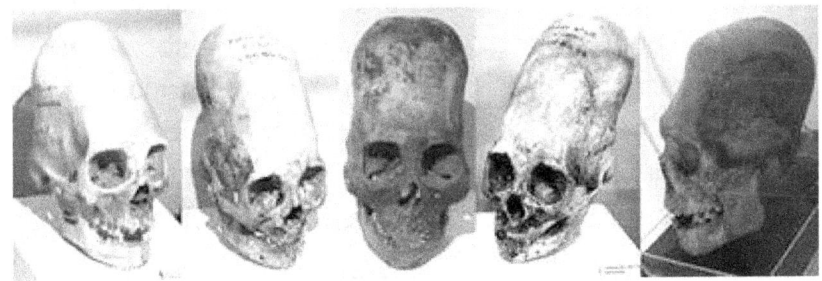

Neanderthal Brain Anomaly

The huge brain of Neanderthal means he was a smart individual. In some respects, he was smarter than today's humans who suffer from brain atrophy.

Some will argue that having a large brain does not necessarily mean the individual is smart. That comment seems to be a crazy assertion, usually made by someone with a small brain or scientists trying to protect their stupid theories.

It's like saying that the dinosaurs were smart because they had tiny brains. Even more than being smart, the difference in shape suggests something else. He had capabilities we do not have nor possibly ever had in our ancient past. I don't know what those were, but the large back portion of the Neanderthal brain suggests that they had to do with enhanced sight, hearing, and motor skills. That's right I said motor skills. He wasn't a lumbering individual and must have been able to perform maneuvers we cannot begin to achieve today with respect to hand-eye coordinated efforts. He had been engineered to be a fantastic worker for the ancient people.

Neanderthal Articulation

According to a large amount of physical evidence, the Neanderthal could punch very clean holes through bones for talismans and they did this remarkable feat with very rudimentary tools. This, by itself, does not show enhanced articulation, but it is a clue. We may also notice that European craftsmanship is highly praised today and the European brains, at least the early ones, seem to be in between Neanderthal and the other "modern" human brains. Perhaps we should infer that a large back portion of the brain denotes enhanced articulation.

Neanderthal Creativity

The other thing that is noticeable is the fact that the "central creativity lobe" is smaller making the brain more flattened than that of our current brains. This suggests that their creativity level was possibly not up to our potential. Whenever the new creation was established, some of these features were not necessary and therefore were not integrated into the design of the man/hybrid. With less creativity came less advancement. Neanderthal made and used only very rudimentary tools even though he was very smart. The two don't seem to go together, but what we will find out is that this hybrid man may not have needed any special tools and his creativity level did not make him investigate newer ways to do things once an adequate method had been reached.

Lack of curiosity evidently killed this cat [I mean Neanderthal].

Neanderthal Worker

These very articulate, highly intelligent, extremely strong workers would have taken more time to develop his society with this lower level of creativity, but that did not mean that he wasn't very good at what he was made for. He was made to be a worker. He worked in huge mining fields that have been found around the world. He worked in the fields; he worked on construction sites; and he worked for the master race of Nephilim until the Adam was created. Then, according to Biblical texts and other documents, within one generation, metalworking, war materials, use of colors, music, and other things were quickly discovered and used. As each new discovery was made by the new Cro-Magnon human, I can hear the Neanderthal saying---"Why didn't I think of that?" The reason was the shape of his brain.

Destined for Extinction

Besides his large, intelligent but uncreative brain, the Neanderthal was a true offshoot with no place to go, [or so it would seem]. Let's look at man's progression pictorially.

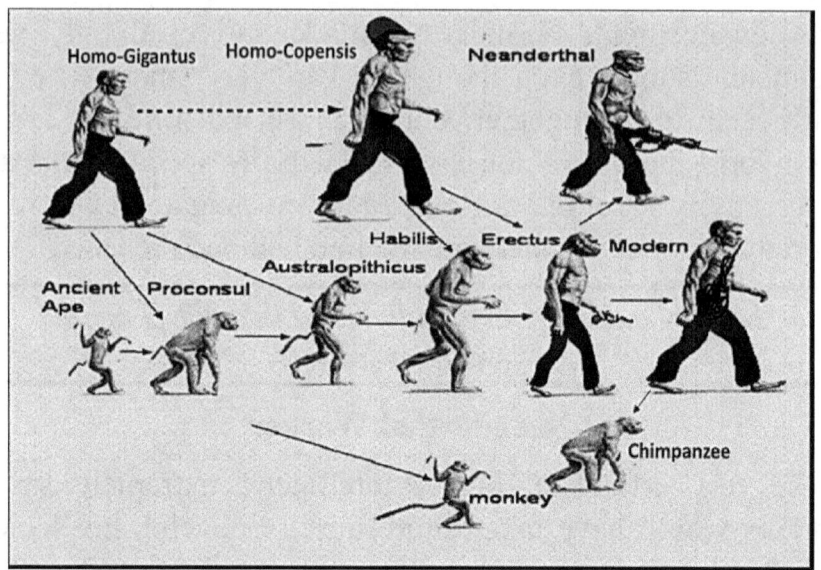

If we take the basic progression of man normally identified and modify it slightly, then the Neanderthal being a combination of Erectus and the giant rulers. The "modern man" indicated in this diagram came from a different line than Neanderthal. You may note the tails in the diagrams. The tailbone of modern humans is much more pointed than that of the Neanderthal. One reason might be that miniature tails were evident on the earlier humanoids, but that has little to do with how smart the Neanderthal were and why scientists don't tell us about Neanderthal.

We might think that because we have 46 pairs of chromosomes, that places us at the pinnacle of evolution and master of the world, but does it?

Chromosome Theory

This seems like a straightforward theory. As creatures evolve, the genetic information contained in chromosomes increases to enhance the species. Wait a minute that is problematic.

This would be easily seen as an increase in chromosome packets; or so it would seem if there were anything to evolution enhancement or uncontrolled evolution in general. On the following page is a short list of common animal types. Beside each animal type is the number of chromosomes used as the building instructions. Notice that "Man" is much more highly evolved than most of the animals as it has more instructions. Wow! The theory works. Man is better than other animals because it is more highly evolved.

Virus	1	Ant	2
Parasitic roundworm	2	Indian deer	6
Fruit fly	8	Mustard	10
roundworm	12	Rye	14
Guinea Pig	16	Dove	16
Corn	20	Horsetail plant	21
Opossum	22	Kidney bean	22
Redwood tree	22	Chinese deer	23
Earthworm	32	Yeast	32
Frog	36	Pig	40
Mouse	40	Wheat	42
Bat	44	**Man**	**46**
Tobacco	48	Apes	48
Sheep	54	Domestic Horse	64
Wild Horse	66	Dog	78
Chicken	78	Carp	104
Crayfish	200	Fern	500
Butterfly	380		

Ape Chromosome Count

Oh! No! It seems that we have de-evolved from the Apes by this logic. Apes, however, have one more pair of chromosomes because two sets of pairs; those called 2p and 2q, are put together in the human set as one chromosome pair, so the theory still holds as our DNA information is actually more compact in humans than in apes.

Apes and the Backward DNA

It should be noted here that almost all of the chromosomes are identical when comparing human and ape sets, so we possibly evolved from them. The only chromosome packets that differ are the 4th and 17th set. These two also are almost identical, but appear to be inverted such that the sequencing is the same but split in the middle and recombined in reverse order. So, an ape is simply an "accidentally backward human" or the reverse with a man being an "accidentally backward Ape". No one knows which came first; the man or the Ape, but evolution is still supreme.

Butterfly Masters

Now we continue down the list and find we will have to continue evolving for some time to get up to the complexity of a dog, carp, or butterfly. We should either respect our master butterflies or disregard this crazy notion of advancement by evolution. Those flying insects know that they are better than us; we don't have to acknowledge it to make it true. Chromosomes acknowledge it for us.

If chromosomes don't indicated evolution what about the penis bone?

No Bone Theory

The human evolution theory continued. It now stated that man came from a monkey in various steps over the last 20 million years. After each cosmic ray bombardment, a quantum change occurred in humans, but something odd happened. Mammals were first generated with a baculum [penis bone] to foster faster procreation. Over time, the bone was not needed simply disappeared as procreation was not as important.

When trying to understand the significance evolution mistakes. This baculum thing keeps coming up. It seems that about 40 thousand years ago there was a marked difference in humans. **The new human lost his baculum**. There were other differences as well. We can believe that this human was the one described as Adam in Moses' Genesis story, chapter 2. Note the tiny tiger baculum to the left as it is important later.

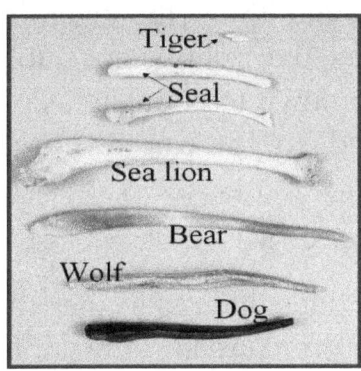
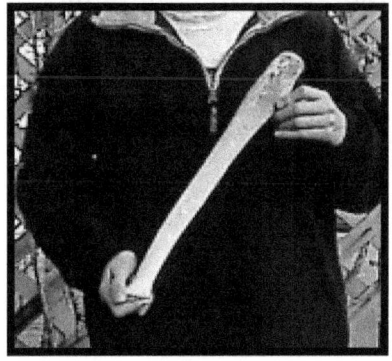

The Walrus baculum is shown to the right. This appears to be where people got the idea of Baseball bats.

> *Another corollary to the bone theory is that Adam, or Cro Magnon lost his baculum when Eve came along.*

I already mentioned that the DNA strains of Neanderthal and modern humans were significantly different and the physical differences of the Homo-Erectus and modern humans is quite noticeable. No Baculum meant procreation was more difficult in the new human. These bones are found in almost all mammals and are specifically designed to allow easy and fast sexual intercourse. The preceding collage shows some of the more popular penis bones. People use these things as charms and keepsakes and, of course the animals used them as well. Without the bone, humans seem to not have evolved from the other mammals. [It should be noted that we have not found a human penis bone for Neanderthal or Homo Erectus, but that doesn't mean they didn't have one of these delicate appendages.] It could also mean that someone has found one and discarded the thing that would mess up his findings.

Size isn't everything. From the penis bones shown at the top in the preceding collage, you would think the tiger is in trouble, but the lion, his close relative with presumably a similar penis bone size, has been known to copulate as often as 225 times within a 4-day period, so you can never really tell.

Modified Adam Story

When we read Genesis 2 we find that what has been interpreted a "rib bone" was taken from Adam to make Eve. There are 2 problems with this fairly common interpretation.

1. Cro Magnon or Adamic man had the same number of rib bones as modern men and women. Possibly, the first

Adamic man was the only one to lose a rib bone, but the next issue is more difficult to ignore.

2. The word interpreted as "rib bone" is not that word at all. It NEVER means that. It actually means "side" or supporting side" or simply "support" and a rib bone cannot do any of these things, but there used to be a bone that was used for support and Adam and all the future Cro-Magnon humans lost it.

Because the Hebrew word could be interpreted as "the entire side of Adam that supports his body", many early images of Eve show Adam's entire side was used to build Eve, but there could have been a different "support member" used.

It has been suggested, that the reason men don't have SUPPORTING penis bones today is that the "support bone" identified in Genesis was pulled out from a much lower area.

Eve had been made from the penis bone so modern men don't have one.

One might wonder about the evolution of humans. Evolutionists would tell you they came as a general characterization with procosul and ape/man joining our ancestor at the very beginning of the Tertiary period [about 100 thousand years ago]. From there, various hops occurred until Homo-Erectus changed just about everything in humans. He doubled his brain capacity, added an opposing thumb, made his feet look like feet, and his hip changed to an upright walking animal within about 30 thousand years. Soon his penis bone was gone 30 thousand years later. Additional Baculum are shown next as the other mammals had a different evolution and kept that bone.

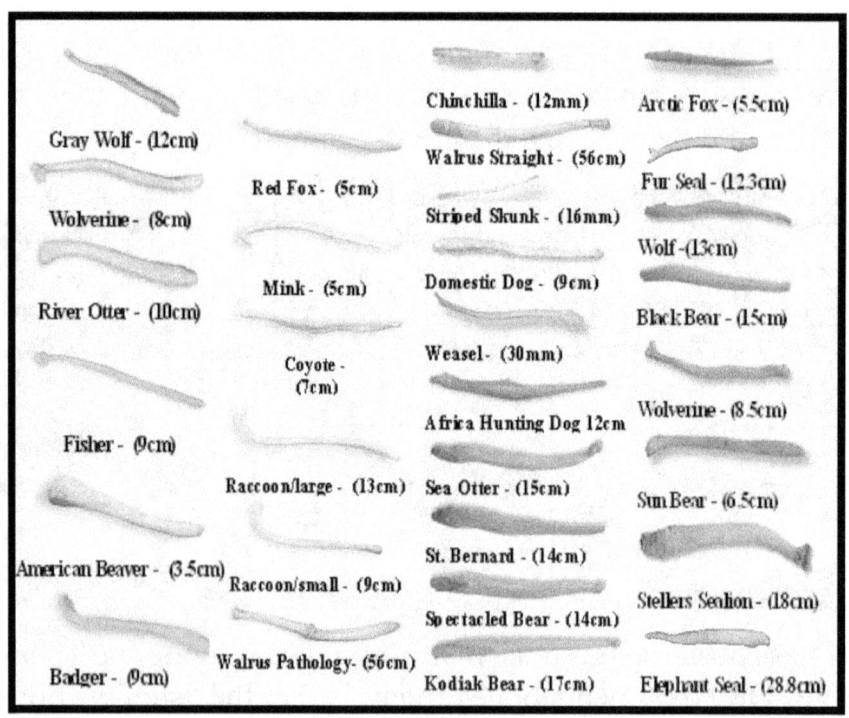

Besides the disappearance of the penis bone, there were massive amounts of change in a very short time. For those questioning the possibility of so many changes in almost no time, scientists got together and changed the evolution theory by adding in cosmic rays. If a bunch of these things hit eggs, they suggested that massive DNA reorganization could occur in hundreds of samples simultaneously and a brand-new species or major change could occur.

If you believe this, I have some Cosmic rays I'll sell you so that your children can become plants or something similar to that. Anthropologists decided if man came from monkeys, they started out as cavemen.

Stone Age Theory

> *This is the silly theory that ancient people lived in caves and grunted while wearing skins of animals from the time that they first came down from the trees until about 6 thousand years ago. They call this time the Stone Age.*

All of a sudden, these Stone Age people began to get smarter about 6 thousand years ago according to some timelines. As I have presented already, ancient civilizations were very advanced in science, medicine and civilization well before the fateful thing that momentarily forced people into caves. Many groups had flying vehicles and world trade was commonplace, but something happened and scientists were too lazy to find out what caused the terrible times. Even after we found out some of the great things that were done before the Stone Age time period there has been horrible resistance to acknowledge that the original theory was totally wrong. Here are a few of the findings.

Sumerian Technology

Sumeria had an impressive list of accomplishments from about 6 to 7 thousand years ago.

Brain Surgery, cataract removal, and bone scraping techniques, fine medical instruments, herbal medicines, beer enemas, and alcohol as a disinfectant, made their medicine fantastic. Models of various organs were used in medical schools.

Artificial gemstones, textile manufacture and basket weaving were common. Lenses, from Turkey to Mesopotamia, there have been found a quantity of about 75 Plano-convex glass lenses. Here's an even more strange part. In at least one of these lenses was found cavities that could have been just to introduce some type of gas or substance to change its optical characteristics. Who knows what they might have been doing with these "modern things".

Use of petroleum products for fire, heat, asphalt road building, and waterproofing were all known. Metallurgical technique included kilns, working bronze, firing metal, casting, and finishing copper mirrors have been found.

Robots, Laser Weapons, and flying machines were all mentioned in ancient texts.

Irrigation, grinding cereal into flour, fermentation, cheese, bread loaves were in use; especially the good old fermentation.

Pottery, ship building and chariot building, and the banking industry showed a high level of society.

The first bicameral Congress, code of laws, and a 4-judge justice system were quickly introduced. The first historians and the invention of writing are attributed. Required schooling and their school used a novel idea; it was called "man in charge of the whip" and there was not as much misbehavior in the classrooms. Large libraries were established [30,000 texts found] and groups of proverb texts. Here are some that are still true today.

Pregnancy Testing- Ancient tablets tell us that insertion of a woolen "tampon" could be removed and treated with "alum". If the wool turned red, the woman was pregnant.

Egyptian Technology

The Egyptian did the same thing. Somehow, 6 to 7 thousand years ago, the Egyptian civilization also went from nothing to a highly technical civilization within a very short time.

Herodotus wrote about the fantastic Egyptian Doctors. He said that each doctor specialized in one type of disease. He also said that Physicians were in every community. Some of these doctors were specialists and would only work on eyes, others the teeth, still others the intestine or internal organs. He also talked about the common use of brain surgery.

The Edwin Smith Papyrus, estimated to be over **5 thousand years old,** has 43 sections just on the treatment of wounds and fractures categorized by body area. Each section contains information on examination, diagnosis, treatment and prognosis of recovery.

The Berlin Papyrus discusses the removal of cataracts from the eye. The papyrus was supposedly found under a statue built during the reign of King Sent who reigned 6 thousand years ago. There are no incantations in the work, as was the custom for later medical documents, so there is secondary cause to believe in its extreme age.

Brain surgery [drilling holes into skulls] was done successfully, just like it was done in Sumeria. They also invented Aspirin and detailed treatment of tumors.

Prosthetic parts have been found; at least wooden toes were made and worn and shows that prosthetics was common and loss of the big toe was common as well. Doesn't this look like the punishment for just about anything was "chop off a toe". Stealing---minus a toe. Winking at a married woman –OFF with your toe! Eating cabbage in on a crowded train—give me a toe and an apology.

As far as chemistry, they must have been very much into that particular science as the name "Chemistry" comes from the original name of the Egyptians which was the Khemites. They had contraceptive jelly and urine pregnancy testing methods. Their mummification was best in the world. OK! It was only best because the desert did most of the work, but they were still good at it.

Molding solid diorite stone [hardest known rock] was evident. Levitation of blocks was apparent. The method was written about and heavy pieces of evidence are everywhere.

Metallurgical techniques like electroplating antimony and gold were accomplished. [most likely with electricity] Models of airplanes have been found suggesting a high level of knowledge about using Flying machines to get around.

There is strong evidence that many types of electric devices were used. Complex irrigation and making beer were both done. I know! What does irrigation have to do with beer? Well, if you irrigate for hops plants you might have to use them on something to drink. Shipbuilding [in the desert] has been evidenced. This was before all of the shaving all of their body hair and crushing beetles and smearing the remains over the eyes to make a mascara so it was not because they were simply nuts.

Great historians and early forms of "complex writing" were established. A society of magic was revered and did things like turning sticks into snakes and making things float. Zep-Tepi, or the beginning of the Egyptian calendar after the "First Time" was approximately 3015BC which coincides with the Tower of Babel coming down. This Zep-Tepi also coincides with the beginnings of the new Stone Age. [It should be noted that the preMaya used the identical date as the beginning of their calendar. Possibly someone should have tried to devise a theory around that similarity.

Dentistry was refined- Gold bands were also used to support good dental operations. Tiny holes were drilled in teeth, the reason for the drilling is unknown.

Chinese Technology

The Far East also was advanced before the Stone Age times. The Chinese ancient society had some of the most extensive knowledge in medicine that must have been passed down from the earliest of times. They also remembered other things.

A Medical Document from before 300 BC stated that some doctor named Chao gave his patients a toxic drink & they were unconscious for three days. Then Chao opened their stomachs up and even examined their hearts. After removing and **exchanging** their organs, he gave his patients another drug and they recovered.

For smallpox inoculations, the Chinese knew to rub dried matter from small pox sores onto the nose of unaffected people to insure immunity. Makes you wonder who was the first to start using the nose crème.

The "Book of Medicine", written about 2650BC, contains this description. It indicated that the blood was know to control the heart and regulate it. It flows in a continuous circle and flows ceaselessly.

Chinese were the first to remember the art of making gunpowder, and paper, and flying, and the compass, and earthquake sensing devices, and many other things which they should not have known how to do given what we think of as normal scientific advancement.

African Technology

African Medicine, Astronomy & Weapons and their knowledge of the universe was unbelievable. In addition to the unbelievable knowledge that the Dogan gained about astronomy, there were other anomalies. Here are some of the more unusual capabilities and elements of knowledge that have been reported from about 6 thousand years ago.

The ancient Dogan tribe knew that the star Sirius had a twin star Sirius B, which was only rediscovered recently. They also knew about two of the moons around Uranus without the aid of a telescope.

A skull was found in Rhodesia. In the skulls are round holes with no sign of radial fracture at the entry point-shattered on the exit. Arrows couldn't cause the holes nor any other weapon besides the high-speed projectile from a gun. The skulls are estimated to be over **40 thousand years old**—well before the time of Samuel Colt and his revolver.

The heart of a ten-year-old girl was, reportedly, placed in the chest of a fifty-year-old man, according an African account from **220 years ago.** It is believed the details of how this should be done came from very ancient medical information passed down for many generations.

Russian Technology

Russians and Mongols also got the "Fast Brain" and knew about complex surgery, architecture and weapons well over **6 thousand years ago**.

A 100-thousand-year-old skeleton showed evidence of Thoracic surgery and healing. I know you have seen many shows about people having sort of thoracic surgery done on them without healing and lots of blood was expelled, but this was substantially different.

A skull from a now extinct Auroch has been found. Like the Rhodesian skull, the small round entry and large exit wounds are definitely those from a high-speed projectile like a bullet.

Indian Technology

From the many books on the subject, you may already know that the Indians knew about flying ships or "vimanas" better than just about any nation, but, not only was it the center for knowledge about flying machines, but also it was a place where many other medical wonders were accomplished and recorded.

The "Samhita", written **8th century BC** contains specification for transplanting tissue. Ancient Indians would rub the crust from smallpox sores on to open cuts to immunize people. Just think what the dark ages would have been like if the Black Death, smallpox, had been cured by this technique or the Chinese technique in Europe a thousand years later.

The "Sushutra" described many surgical techniques including removal of neck tumors, Tonsillectomy, and amputation techniques. It describes proper use of over 121 surgical tools and had a detailed description listing of many diseases of the day.

"Celus" confirmed ancient transplanting techniques used by the ancient Indians during Roman times. Transplanting skin to manufacture a new nose whenever one was cut off in battle included partially cutting off portions of the skin from an arm and attaching the skin to the face while still attached to the arm. This allowed the graft to retain blood-flow during healing.

There can be little doubt that societies were thriving and highly civilized well before the horrible times known as the Stone Age. It was not the beginning of mankind, but the remains of some horrible event in man's history that shoved us back into a Stone Age condition.

A Better Answer

The Biblical book of "Jasher" tells us that during the time of the great Tower of Babel there was a horrible war which took the lives of 1/3 the population of the world and thrust another 1/3 of the population into a life like an ape. About this same time frame, we find the remains of what has been known as the most advanced ancient city, Mohen-jo-Daro "mound of the Dead". Strewn all over the streets were the skeletal remains of the dead and globs of glass which had once been bottles and jars, all melted from some intense heat. Like the remade dinosaurs of the Pleistocene, the bodies are still radioactive from whatever destroyed that city and just about everywhere else around the world. The pictures below show the remains of the city and its occupants. The only surviving civilizations, apparently, went underground as dozens of underground cities have now been found in Turkey, Malta, South America, the United State, Scotland, Mongolia and other sites. Whatever happened was massive and deadly so don't let anyone simply tell you that the first men lived in caves 6 thousand years ago.

 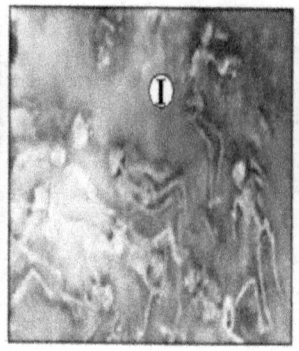 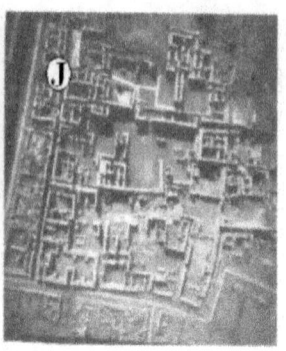

It is true that one of the outcomes of the war was we lost the ability to use much of our brain and it began to atr4ophy making it 20% small, now, that the earlier Cro-Magnon people from before the war. Loss of memory destroyed civilizations and for a brief time, man reverted to a much simpler life about 5 thousand years ago. We might call this regression a new beginning as the Egyptians, PreMaya, and Indians describe, but there would have been fire making, metal working, vocalizing, and women's liberation so a Stone Age is way off base.

Physics Misconceptions

Big Bang Anomalies

> *This theory essentially says that there was this massive chunk of matter in the center of the universe before there was a universe. The mass, sort or, got pressurized and burst spewing stars and solar systems and galaxies and everything else that makes up our universe.*

Everyone has heard about this thing and our schools preach the gospel of Big Bang. You aren't even allowed to present something different. The travesty is that it is all a fake. At best, it is a mathematical shortcut to allow us to define a beginning. It is nothing more. "Scientific Creation" and reason have problems with this major "universe beginning" theory, which indicates that 15 billion years ago the universe was created, <u>without help</u>, in a huge explosion or "big bang". Research into the details of the theory has shown it to be terribly flawed. That doesn't mean that many of the components of the theory have no relevance and that there is another theory that provides all the answers because there is not; at least as far as I can tell. What I want to show here is that people take these theories as TOTAL TRUTH. If the facts don't meet the theory, they figure that the facts must be wrong rather than the other direction. Here is a list of the various components in dispute with regard to this, almost universally accepted, but substantially flawed, conclusion about spontaneous eruption of our universe.

Limited Matter Anomaly-In the universe, there is only 1/100th the amount of matter needed to support the Big Bang according to W.H. Mcrea, *"Science and Creation"*, 1971.

Stellar Rotation Anomaly-The Stellar rotation is too fast to support the big bang theory according to Richard Johnson, *"No Way Out"*, 1963.

Antimatter Anomaly-If matter was created during the big bang, an equal amount of antimatter must also have been created, and there is almost none according to R.M. Somerville, *"Cosmic Mysteries"*, 1990.

Lumpy Universe Anomaly-The law of entropy indicates that the universe should seek its most random condition after this long time from the "Big Bang". Instead, lumps of galaxies are all over the place according to P. Peebles, "Science", 1982, and Guth and Slunhart "Scientific American", 1984.

Low Radiation Anomaly-Too little radiation has been recorded from the stellar masses. This does not support a growing universe model as indicated by the big bang theory- according to William Corless, *"Stars, Galaxy, Cosmos"*, 1982.

Red-shift Anomalies-The Big Bang was dismantled by direct observation-including a highly red-shifted quasar in front of a nearby galaxy! For many years it has been known that the map of the universe acquires a bizarre appearance when you let red-shift determine distances. Suddenly galactic clusters stretch out in radial lines absurdly pointing **at the earth**. The effect is called "the fingers of God," and the earth-directed "fingers" span billions of light-years. If all these stars look like fingers going away from us, then the earth is in the center of the universe where all of this big banging took place.

In addition to the fingers of God thing, video spectrum red-shifts of quasars are too great. These red-shifts indicate that we are able to view stars that are almost 15 billion years old and at unbelievable distances away when using the big bang model. The stars would have to have been larger, brighter, and denser that anything that we could possibly imagine. The problem is that **several** of these supposedly immense stars have been located. They all have the same anomaly and every one is unbelievable, according to *"Time-Life"*, <u>Stars,</u> 1988.

Let's think about the whole "fingers of God" thing. IF the earth were the center of the explosion, people wouldn't exist and the Earth wouldn't exist anymore.

As a model, I believe that something like a Big Bang actually did occur and it is a good way to describe what happened, but there must have been some outside force that helped it happen and shaped its outcome, there must be some characteristic change associated with consciousness as discovered in the Theory of Relativity, and the theory should be addressed as a problematic hypothesis that people should question.

The Big Bang was immediately converted to the Ekpyrotic Membrane Theory which required both Heaven and our universe to be "created" simultaneously and the Heaven Universe continuously replenishing matter, energy and life itself and they all escape our universe to feed the Heaven universe. The Big Bang had begun. One scientist made a bode law that placed our planets firmly in position, The problem ws it was all bogus.

Bode's Bogus Law

The law defines the distance away from the sun each planet revolves by a strange menagerie of additions, and doubling from a minimum number of 3.

Using the earth distance as 10, for no particular reason as 10 equals 93 million miles so 1 equals 9.8 million miles. From that craziness, the theory/ LAW uses a base number of three that is doubled for each planet distance before the number has a 4 added to it. The LAW continues its bizarreness as the outer planets use a doubling of distance from the previous planet. I hope that was as hard to follow as it was to write.

People had been trying to force fit the planet positions into a neat package for years without success. Finally, in 1766, this theory was invented by a man named Johann "Titus", so we got the bright idea to call it Bode's Law. It is not even close to being a law, but it had a ring to it so people kept calling it that. The law used the outrageous mathematical formula for holding planets to their current orbits that I stated above. Some of the planets seem to be close, but when it comes down to it, the positions don't match. Our Solar system is still forming. I know it's hard to believe, but even as recently as 12 thousand years ago there appears to have been a major repositioning of the Bode's Law controlled masses. The best thing we can do is to simply ignore it. Cross it out of your science books. Here are some of the time-stamped diagrams of our solar system through the ages. These don't fit the

Bode's Law Theory, so if you are a die-hard, please skip this section.

Correction to the Law to Account for Anomalies

We are told that the solar system generally was established 4.55 Billion years ago and with it, the planet Earth. New evidence suggests that there was more to it than that. Many of the planets did assume their place, but a planet close to Mars was orbiting and no Earth [as we know it] was evident in this time slot. By the way, thousands of planetoids sort of jumbled around and essentially orbited the sun in what is called the Oort cloud just like they do today.

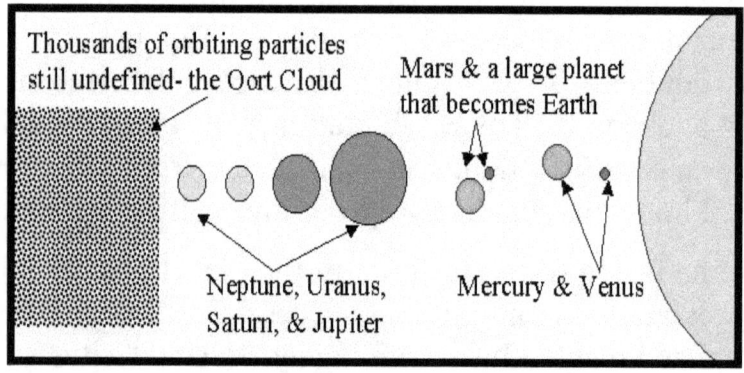

Solar System Changed

Current evidence strongly suggests that the Solar System changed about a million years ago. The planet close to Mars finally exploded from the gravitational pressures and the asteroid belt is established. The largest part became the Earth. I know that concept seems fanciful but please continue reading and its probability will become more understood.

Beyond not believing this event occurred ourselves, we have a very difficult time believing that this event would be understood by ancient man, but many historical references are almost identical in the details of this event and new computer simulations of this event help confirm some anomalous characteristics of the solar system of today. You may not believe this fact right now, but the proof is provided in later chapters.

Oort Cloud and the Kuiper Belt

The Oort cloud also split around this same time frame. At least 31 of the planetoids established a somewhat more stable orbit around the sun. The more stable area is now called the Kuiper belt. The largest planetoid in the group is named Pluto. Exactly how it got there is unknown. According to interpretations of the Sumerian texts by Zecharia Sitchin, Pluto was a moon that was pulled away from one of the Jovian planets as another extremely large planet entered the "known" Solar System from its eccentric orbit in the Oort cloud. I'm not getting into additional planet theories in this work, but certainly it is possible to have large masses with eccentric orbits in the Oort cloud. These masses could take thousands of years to orbit the sun.

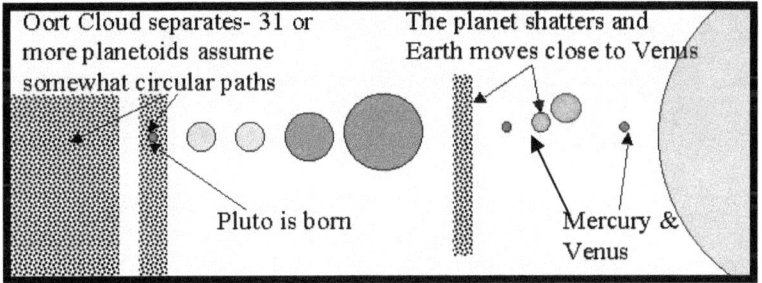

The Solar System Changed Again

About 12 thousand years ago the evidence suggests that another interesting change occurred. The moon of Venus exploded which sent thousands of particles raining down on the inhabitants of the Earth and set up Venus for a terrible disaster from which it never recovered.

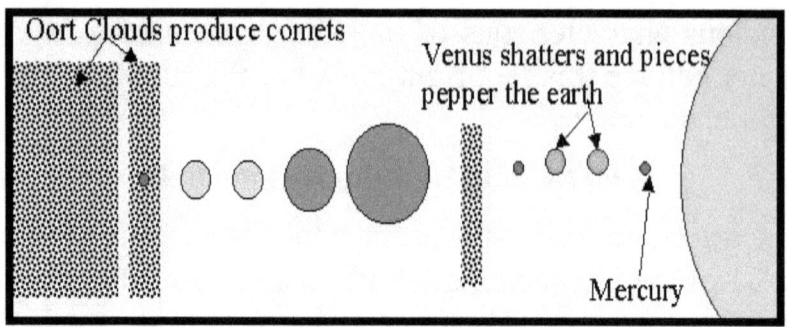

Even when scientists know Bode's Law is stupid, they don't examine other probabilities as they disrupt what the scientists WANT TO BELIVE. All these wobbly planets live in what we call the universe.

The Universe Anomaly

The new understanding is that the universe is endless and made up of vibrating nothingness or quantum fluctuations and it must be linked to a second universe that continuously replaces everything that is lost every single second.

Has anyone told you the absurd suggestion that the dimensions of the universe are length, height, width, and time? While this makes no sense anymore as matter doesn't "actually" exist, our students are being bombarded with this stupidity all the time. After all, "Its written in a book!" Even with all the newfangled theories that attempt to correct what is being shoved down student's throats, when someone mentions "dimensions" we should say something like Life fluctuations, particle fluctuations, and force fluctuations to come closer to a truth.

If we are near the center of the universe and it is endless, everything should be dark. Ignoring this distraction, today, scientists indicate that everything is made up of vibrating nothingnesses. These are called "quantum fluctuations" to hide the fact that they are talking about nothingness. Light is made up of vibrating electrical potential, matter is made up of vibrating Aether, and Life potential surrounds DNA as a vibrating nothingness as well--- I mean quantum fluctuation. According to Einstein, if you turn on a flashlight and photons are thrust out at the speed of light, energy is moving away from you. No matter where you point it, energy is moving away from you. Soon, the photons are so far away from your location, we would not be able to use the energy. Even the light from the stars would mostly travel towards the "end" of the universe.

Here is the problem!!!! Energy, matter, and life cannot be created, according to many sciences so after a time all energy, life, and matter will be gone. As light and energy passes away from the various objects, the energy would slowly escape to the ends of the universe, never to be seen again. Soon there would be no more universe as there would be no more matter. It would have all escaped. Now for our savior.

Heaven Universe

A second universe is always linked to a primary universe so that as energy, particles and life fluctuations all reach the end of one universe, they are immediately replenished from a second universe that just received the fluctuating emissions of our universe. Many call this linked universe Heaven, so I will do the same. Without heaven, our universe would die as all the length width and height stuff would disappear. Because some don't like to talk about Heaven being real, our books have still not been changed and our students have very little information to allow them to understand our universe.

Besides not understanding our universe, we hide issues with how the Pacific and the mountain range that surrounds it was made. I can only imagine the reason is to completely confuse children so they won't ask questions.

Earth Misconceptions

Plate Tectonic Theory

This theory simply states that massive plates of the earth got moving around one day and rammed into each other so hard that the plates were pushed into the are over 2 miles. For at least one of the mountain ranges this push up extended over 15 thousand miles, over half way around the world.

Here is one where a small amount of scientific fact is mixed with a whole lot of fantasy. People felt that they would be discarding the fact if they didn't believe the fiction, I suppose, so the plate motion, mountain making, tectonic theory was born. For a minute let's go back to school. From the time you were in grammar school, all the way through college, you were continuously told that "plate tectonics" formed the mountains of the Himalayas, reasonably called the Himalayan Ridge. This same thing caused the long range of mountains called the American Ridge that cover the western side of South and North America. We were told so many times, that we didn't question its absurdity. On the next page is a graphic of the idea. The plates simply went past each other and all the dirt and gravel that was on top built the mountains

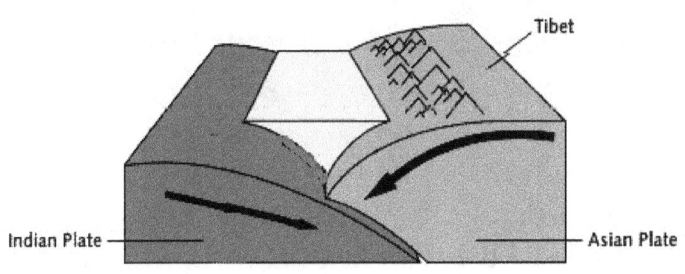

If the Himalayan ridge is the intersection of two "Plates" and one plate was rammed against another huge mass about 200 million years ago such that **"quadrillions of tons of Earth"** were pushed up an average **"2 miles into the sky"**, it took more than a few little volcanoes to push the slab with the force required. It would have taken a cataclysmic event we can't even imagine. Then, supposedly, the same thing happened again for the American mountain ridge.

Estimates have been made that **the two "volcanoes" required to do this would each have been about the size of the United States**. There just isn't any evidence of these huge explosions. The whole concept is so ridiculous now, that I'm embarrassed I ever believed the tale in the first place. The problem is the same as that described in the earlier illustration with the eye-leg similarity. Like the other, the science community started with fact. Plate tectonics has been proven and evidence of its existence is seen every day. Then they went berserk because they couldn't figure out what caused these huge piles of rock. Instead of letting our children know that there are **"unknowns that should be explored"**, we continuously insulate, stagnate, and destroy the minds of our children. We sort of liquefy their thought process so that it can be molded into thinking the way some group of our society wishes people would think.

Force Vector Evidence

The early theories that mountains were pushed up by plates moving together are not only discredited by size but also by direction. People began to wonder why the mountains all fell in straight-line patterns, for instance. One path is straight along the equatorial line and another is even stranger as it goes along the side of the Americas and along the same path on the other side of the world. ***Plate tectonic models cannot go around more than 50 percent of the globe***, because there would be no way to push the block, but the tectonic theorists had to make them that way to support the mountain ranges until sanity finally won out. OK sanity never did win out, but we can always hope. A diagram of the American Mountain range along the border of Pangea is shown next as well as the direction of the force required to build the range. It's pretty much ridiculous. I suppose one could say a hundred volcanoes erupted simultaneously along some group of plates to begin the massive move, but what would have caused all the volcanoes?

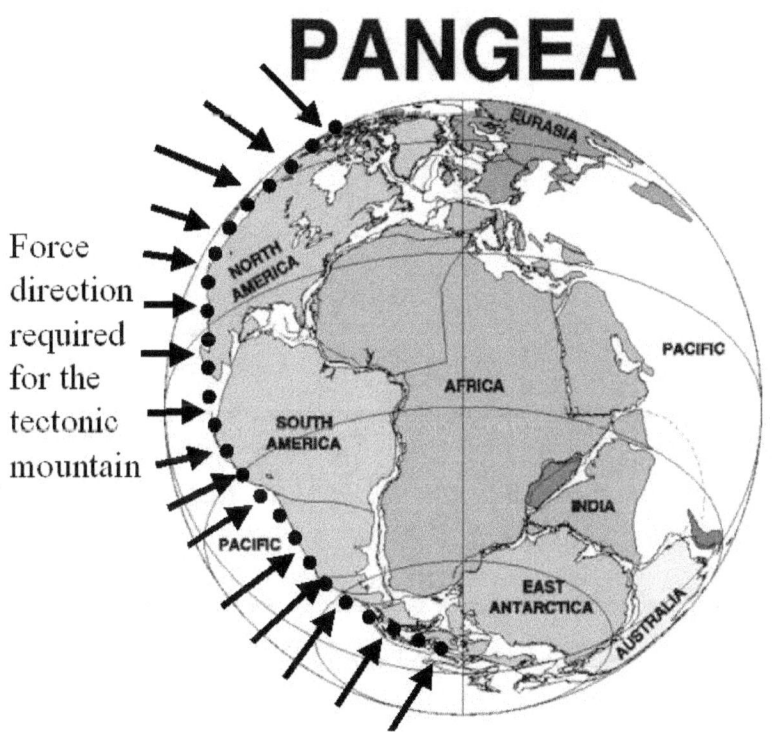

There might be a more palatable theory that won't make your brain hurt.

Uplifted Mountains Theory

As a corollary to the previous. This one says that was all bogus, but here is a possibility, as data suggests another planet came so close to the Earth that it pulled up the mountain ranges.

I know you had a little chuckle starting, but listen to the evidence. The much less plausible push-up theory using plate tectonics is always used to explain how mountains were formed. The uplift mountain theory is a much better one that actually works when the old plate tectonics one falls flat on its face. It is impossible for the major ranges to have been formed that way, but never fear, plate tectonic mountains will be taught in schools for a long time to come.

After you get a little more detail, this, seemingly impossible, theory shows one of the only reasonable methods to produce the long mountain ranges. The most likely method was an encounter of another planet. As the planet came close to the orbiting Earth, the gravitational attraction between the 2 was tremendous and the land literally was pulled out to the long mountainous regions of the world. The one we should look at in particular is the American Range that starts at Antarctica, goes along the western coastline of both Americas then travels around the other side of the world to Russia. The mountain is in a straight line for one simple reason. The earth was spinning along the axis perpendicular to the mountain range at the time and the range was created along the "equator".

Zoroastrian History Evidence

We may get better insight from the ancient Zoroastrians. Here is what they had to say about the formation of the mountains.

*"As the evil spirit rushed in, the earth shook, and the substance of mountains was created in the earth. First, Mount Alburz **arose**; afterwards, the other ranges of mountains (kofaniha) of the middle of the earth. (arose).."*
[Some outside force created mountains around the middle of the Earth.]

Thanks to many satellite pictures and computer models, we now know that the Zoroastrians were generally correct. The Mountain ranges actually formed along the equatorial perimeter during some cataclysmic event and the mountains were pulled up into placed rather than having two dirt piles pushed together.

Energy Evidence

A huge amount of energy was expended to cause the earth to split open like this and yet there is absolutely no evidence that this huge amount of energy was enough force to push the earth's crustal plates hard enough to form mountain ranges and that brings us to the bogus "Plate Tectonic" theory again. If that force wasn't enough to push two of the plates fast enough to cause the Himalayas just what was supposed to have caused the great plate smashes? Schoolbooks typically don't cover these huge lava flows because someone might ask that very question. They also don't talk about the Earth splitting open because someone might get uncomfortable when thinking that the Earth could possibly come apart. Unfortunately, it could and being comfortable doesn't change reality, my friend. One of the reality things" is that the earth's rotational axis is STILL unstable.

Reasonableness Without Plate Tectonics

With this knowledge, the most likely culprit was easily determined. A large planetary object strafing the planet made each of the extended mountain ranges. Once the Earth was strafed with the Earth rotating on an axis through the middle of the Pacific Ocean and Asia we can be comforted in our theory as it happened a second time when the rotational axis had shifted and was similar to our present rotational feature. Yes, I did say the Earth's axis changed. In fact, we will see that it has happened more than once. The picture below shows the projected path of the planets close encounter on two separate fly-bys. The wide lines represent the long strings of mountains along the 2 paths. One going from the southern tip of South America through the tip of Alaska and down along the coast of the Far Eastern countries. The second, more severe uplift included the region from the Middle East through Pakistan, Tibet and China.

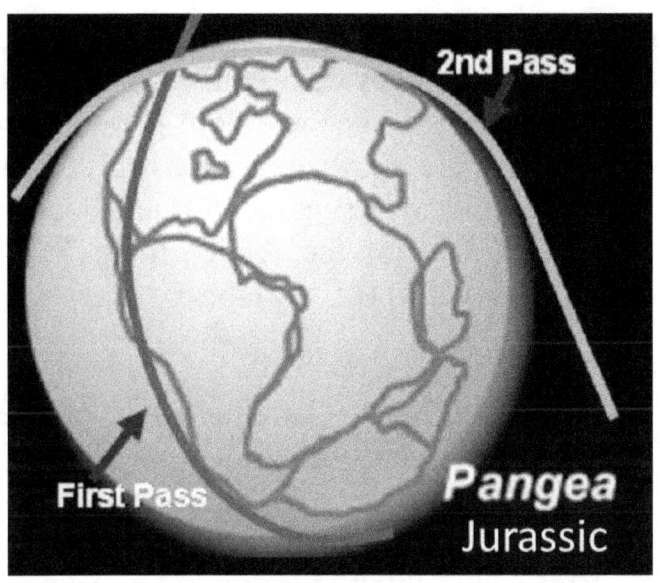

Pangea Reasonableness

Actually, the Planet flybys occurred prior to the time when the Pacific Ocean was formed and before the continents began to move which finally began to fill in the Pacific hole. The picture shows potential "Flybys" as described, with all of today's continents together as one clump. This clump we typically call "Pangea" the super continent. On the other side of the Earth would have been a similar super continent we can call Presotonia. Prestonia allowed the earth to rotate without wobbling from the Pangea bulge

Another Curious Point

A huge amount of energy was expended to cause the earth to split open like this and yet there is absolutely no evidence that this huge amount of energy was enough force to push the earth's crustal plates hard enough to form mountain ranges like you were told in the bogus "Plate Tectonic" Theory. If that force wasn't enough to push two of the plates fast enough to cause the Himalayas just what was supposed to have caused the great plate smashes? Schoolbooks typically don't cover these huge lava flows because someone might ask that very question. They also don't talk about the Earth splitting open because someone might get uncomfortable when thinking that the Earth could possibly come apart. Unfortunately, it could and being comfortable doesn't change reality, my friend. One of the reality things" is that the earth's rotational axis is STILL unstable.

Clams

Isn't it odd to find giant clams on top of the Andes Mountains and schools don't bring them up? Plate Tectonics is not the answer, by the way. If this concept were true, all this uplifting surely would bring the ocean up into the mountains. You would think there would be some evidence of mountains rising high in the air during this time period and there is. Before the first "uplift" from a Mars encounter [Forget I said Mars as I'm not getting into all of that in this book but Mars was almost destroyed about the same time that our Pacific Ocean hole was made.], the western side of what is now South America was underwater rather than being a huge mountain range. Today, evidence of the top of the mountain range being underwater can be readily seen. In the following graphic, piles and piles of HUGE prehistoric clams, the size of a man, have a man sitting on them. This is on top of the Andes Mountains in Peru. If the land and the clams weren't pulled out of the water together, I don't know what allowed the calms to be there

Timing The Mountain Range

Scientists even know when the mountain range was produced and the clams were deposited, but they are not telling you. It was the end of the Permian during the worst extinction of all time on Earth. Let's first look at Ice core samples from Antarctic. What we find is that every extinction is neatly shown as a massive increase in temperature. The end of the Permian 440 thousand years ago the Earth's cyclic variation changes completely as if the planet changed drastically.

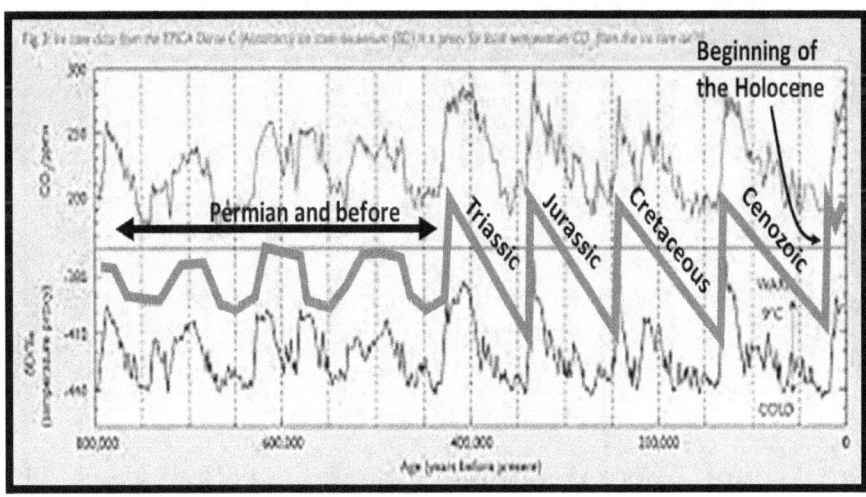

It is as if the planet wobbles more now that there is a hole that would become the Pacific Ocean on one side. Please don't worry that the timing seems shorter than what you were told in school. We will look at that more later.

Pangea Split Timing

Sometime after the North American-South American mountain range was made, a big chunk of our planet was ripped away and the hole is now the Pacific Ocean. Pangea became the only remaining continental mass and it began to split apart to fill the rupture. The "filling-in" is still occurring today, as the Atlantic Ocean gets wider. According some models this split started at about 300 feet per year and slows to about 2 inches per year today making a current average width of about 6,000 kilometers. Each time the separation is enough, the split at the bottom of the Ocean starts spewing magma make the mid-Atlantic ridge shown in the next graphic. From this data we can determine that the close encounter and the explosion that made the Pacific Ocean would have occurred mathematically, around 440 thousand years ago just as the Ice described.

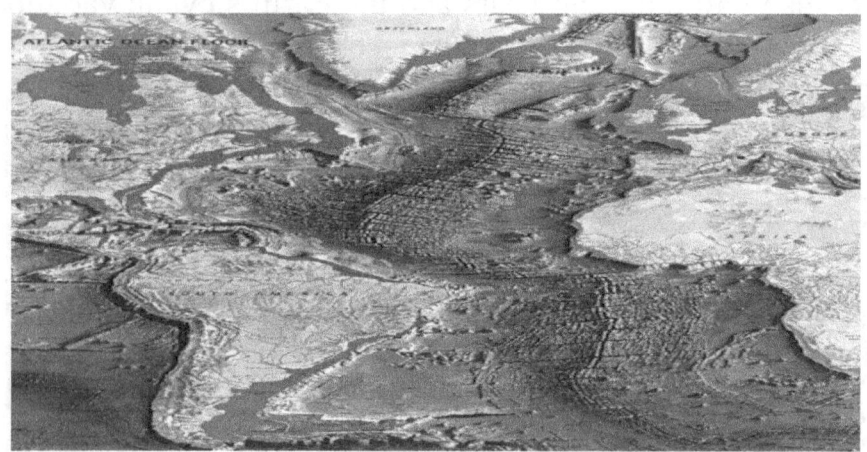

Pacific Ocean Facts

The Trough-There is a trough around the entire Ocean. This type of ditch is not indicative of land massed pushing together. Instead it shows where Prestonia had been yanked out.

Crust-The crust under the Pacific Ocean is almost non-existent while the crustal depth under all other oceans is the same as that on the land areas. This shows that the crustal mass at this location was yanked away.

Theories continued to expand and books like *"The Biblical Flood and the Earth Epic"* with editions in 1966, 67, 68, 69, 70, and 1971 continued to tell one of several versions of this horrible event that wasn't based on fantasy like that which is pushed by the Vain scientists.

Trench-The Mariana's Trench is along the same direction as the Ocean trough showing that the trench was an artifact of whatever happened to the land above the Ocean.

On Mars we find Similar Details

No Craters-The Martian planet is a strange one. The entire Northern half of the planet has almost no huge crates while the southern half is completely covered. This indicates that the planet was ripped in half not very long ago. [possibly 440 thousand years ago] This is shown in the following topographical image of Mars.

No Crust-The crustal depth on the northern portion of the planet Mars is almost non-existent indicating that the planet was ripped in half at one time. This can also be seen in the following image.

Trench- Like the Mariana's, a huge trench the length of the entire United States is along the separation line between the no crust and crusted portions of Mars. This can be seen in the next image. I have circled the area. It is called the Valles Marineris.

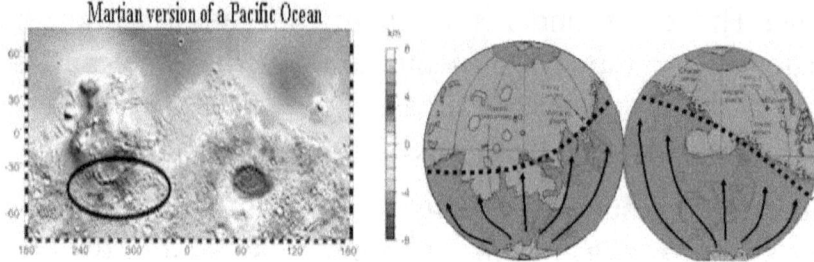

Pangea- While we are told that Pangea was the only super continent of the ancient times, this would be impossible as the earth would tend to have a stable land mass. We can presume that if a massive continent was on one side of the earth, an similar one would have been on the opposite side. This continent is gone so we can assume that something yanked it into outer space.

Asteroids- No one knows where the asteroids came from except that they are planetary masses. One theory is that a planet used to be in the orbital area of the asteroids and is was completely annihilated. That is pretty much impossible so the material probably came for planets that are still here. Two that are nearby are Mars and Earth.

Our Moon- Our Moon is too big if it were a simple grab of material that came near our planet. A much more reasonable place to have come from was the Pacific Ocean.

Revolution- The Moon's revolution being in sync with the rotation tells us the Moon, most likely, came from the earth.

No Core Mass- The moon has almost no core mass or metallic structure suggesting that it was made up of a continent ripped away from the earth such that the heavy core stayed behind.

Mars Got the Worst- If Mars and Earth came close together and the gravitational pulls of both planets reacted with each other, Mars would certainly have gotten the most material pulled away. Let's say ½ the planet would have been pulled away and the entire northern hemisphere would be missing.

Mars Healing- The evidence shows that the southern hemispheric continental mass is slowly going towards the northern hemisphere suggesting that once, the northern hemisphere and southern hemispheres both had similar structure and the northern area lost something.

The same healing process is evident on earth as a massive continent was now splattered all over the solar system. Pangea had been on one side of the Earth and Prestonia was on the other. Prestonia was sucked into space, so Pangea began filling in the hole.

Missing Continent Theory

Common Theory is that the Earth looked around one day and the Pacific Ocean hole appeared and started filling with water and Pangea split apart to try to fill in the hole. Anyone telling you about the supercontinent of Prestonia is a charlatan. A corollary is the continent of India had been an Island and it rammed into Asia making the Himalayas.

As I said in the last section; the Earth lost a passive amount of earth at the end of the Permian Age. While vain scientists keep trying to hide the manufacture of the Pacific Ocean, the truth is well known. One day, while minding its own business, something came along and ripped ½ of the entire crust off the planet and it could happen again so we should, at least know what happened, why it happened, and when it happened so we can be prepared if we have to be. Even after hundreds of thousands of years and after the Atlantic Ocean opened up to try to seal the hole, we can see the huge loss of crustal surface just below the water.

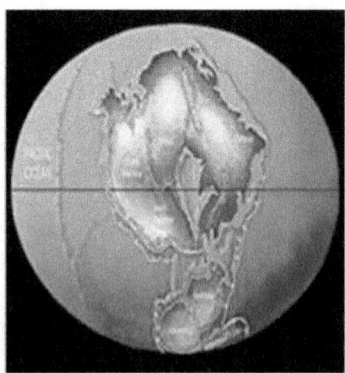

Now there is a huge hole in our planet and no want wants to tell our students that a huge mass was <u>sucked out of the Earth</u> so it is somehow ignored. It doesn't mean it didn't happen, it simply means we lie to our children because we don't like the truth and we think that ignorance is better than being uncomfortable. Below is a topographical cross-section of the 24,000 mile long earth crust as it is today [across the Tropic of Cancer].

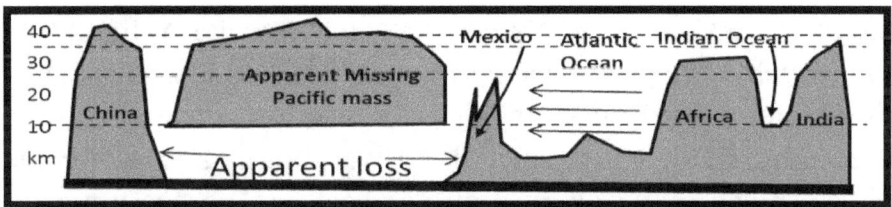

If that was scary enough, the graph below shows that the loss was much worse than it even looks today as <u>at least ½ of the Earth's crustal mass was somehow yanked out</u> in space.

A google Earth view of the Earth today, centered at the center of the hole even after Pangea has been yanked apart to try to heal the massive hole still shows how very horrible this event was. See next left.

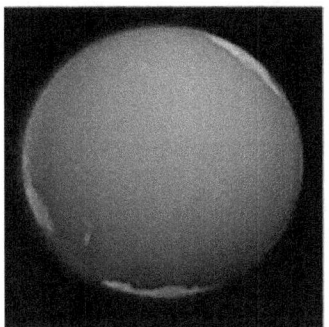 

There is almost no crust below. It's all gone! Teachers and vain scientists teach a lie by not providing the details. Other works of fiction forced down the throats of children concerning this horror are as following:

Plate Tectonics formed the massive mountain ring around the Pacific Ocean. The naturally formed ocean plates were slammed in every direction at the same time pushing the ocean rim over 5 miles in the air. While there is no trace of the massive volcano or plate pusher in the center of the ocean to do such a thing don't ask me questions about what pushed the plates. The missing plate pusher is shown on the preceding right. Of course, there was nothing to make the plates smash outwards in a circle. Tectonic Forces typically work in only one direction, but it didn't fit the "clean version" of our planetary development so this fanciful expanding ocean stupidity is still in the science books today.

After going over this twice, I hope it is making more sense; if not, please use the plate tectonic theory from school. While you are at it, continue to use nuclear timing if you don't care about truth.

Nuclear Timing Disaster

Some still indicate Nuclear Timing is constant and all events can be accurately timed by this fantastic standard.

What I'm talking about here is nuclear decay as a form of "accurate dating". Initially, nuclear decay was tested to see how consistent it was and it seemed pretty accurate so everyone got on the bandwagon of nuclear decay. Vain scientists could just look at percentages of nuclear isotope in a substance and instantly know how old it was. We could even time the earth with something called Radioactive Argon testing.

Oh! Happy days! Oh! happy days!

For some time now, <u>ALL</u> have known the huge issues in this previously established timing baseline. You have been told dinosaurs died 65 million years ago and the beginning of the Mesozoic Era was 300 Million years ago. You were told, over and over and over again. They even proved to you that was truth by telling you lead, potassium, and even carbon isotopes decayed at a set rate; just see how much of an early isotope is left and read the date. Besides, the dinosaurs are buried underground and turned into stone so there had to be a long time for that to happen.

While the earth is ancient, it is definitely not as old as has been told to you. Many of geologists today still tell you that radiometric dating has narrowed the age of Earth to about 4.5 billion years, give or take a couple of percent. We now know that the dating method is inaccurate and scientists actually desiring to find truth are refining the timing more and more each day. The Earth and everything in it is much younger and so are the characteristic stabilities of the planets in our Solar System. Researchers at Purdue and Stanford have found evidence that-

Radioactive decay rates are not constant at al with an error rate of about 5000% *Don't tell our children, don't modify any science books, don't make statements on Television, don't completely retime the earth and all of the various ages we were taught in school. That would disrupt our history, science, paleontology, evolution, Astronomy, and just about everything. It would be much better to leave it alone. [the vain quasi-scientists said]*

On December 13, 2006, a magnificent solar flare flung radiation and solar particles toward Earth. Measuring the decay rate of manganese-54 during the flare proved to be very interesting as the decay rate dropped during the time of the radiation fallout. It was determined that solar neutrinos zipped through space and affected Mn-54's decay rates used in the experiment. Just think about this. They were testing a single solar flare event and the change was significant. The sun has these things all the time. It was also found that the decay rates of silicon-32 and radium-226 showed seasonal variation, according to data collected at Brookhaven National Laboratory on Long Island and the Federal Physical and Technical Institute in Germany. This error was just the material sitting there with almost no outside interference. Wood buried in igneous rock in Queensland Australia has been dated to 40 thousand years, while the basalt around it dated to 45 million years. Both dating subjects should have given the same date, since the igneous rock was formed at the same time the wood was buried. Many of the "data-ologists" don't tell you about major errors like this.

Lava Errors- Excess argon-36 was found in three out of 26 lava flows in recent times. So <u>Argon/argon testing would show a much older date that actually was "KNOWN"</u>. This is believed to be because there was too much of the argon-36 in the first place. In the Grand Canyon lava flow testing showed lower levels of lava was younger than the top layers. At different volcano sites, that had eruption in 1949, 1954 and 1975. The same thing was noted. Geochron Laboratories of Cambridge, Massachusetts dated these samples. Even though the <u>oldest of these samples are just over sixty-years old</u>, the lab tests provided ages that <u>ranged from 270,000 years to 3.5 million years old</u>. Additionally, we go to Mt. St. Helens and its eruptions in the 1980's. Samples there gave old ages in the range of <u>300,000 to 2.7 million</u> years. Hopefully, you are beginning to see that we know less about how old we are than you believed before reading this.

If neutrinos from a single solar flare can make things look older, what if the entire Earth was closer to the sun?

I know that sounds odd, so just keep it in the back of your mind right now as we try to find some standard for dating.

Nuclear Decay a Bad Timing Method

Today we know that the nuclear decay dating of things including Electron Spin Dating and Uranium Dating, Thorium Protactinium Dating, Oxygen Sediment Dating, Lead-lead-lead Dating, and Argon Dating [which we originally used to date the ages of the Earth] are terribly flawed. The old standard carbon 14 dating also seemed in jeopardy. Dating beyond about 30 thousand years was <u>much younger</u> than tested. If there had been nuclear events [bombs or even volcanic eruptions] the apparent timing was changed drastically. Other methods had to be employed to determine how everything should be timed, but classroom information was not changed. That would confuse the students. I'm going to prove to you how you have been lied to by vain quasi-scientists. This will give you a better understanding of the lengths some will go to when they believe something, no matter what the evidence shows.

Standard Geological Timeline		
Era/Period/Epoch	Time (M yrs. ago)	Time (T yrs. ago)
Archaeozoic Period	5000-1500	50,000-3000
Proterozoic Period	1500-545	3000-1000
Cambrian period	550-500	1000-900
Ordovician period	500-440	900-800
Silurian period	440-410	800-700
Devonian period	410-365	700-600
Carboniferous	365-300	600-500
Permian period	300-250	500-400
Triassic period	250-212	400-300
Jurassic period	212-145	300-200
Cretaceous period	145-65	200-100
Tertiary period	65-1.8	100-40
Pleistocene period	1.8-0.01	40-10
Holocene period	0.01-0	10-00

The middle listing of dates above is the "STANDARD" that had been presented in our classrooms, while the last column shows a somewhat closer, more accurate time line that has been verified by MANY, MANY non-nuclear decay methods. Even with the mountain of evidence showing how nuclear decay cannot be used, the middle timing is still heralded as the master in many schools and books being used to teach our children without basis just like most of the other stuff we have been talking about. I know it is difficult to believe historians, scientists and teachers would keep these things from you, like how does greenhouse gas affect our planet, so let me tell you a little more.

Stratigraphic Position Timing Issue -Besides Nuclear decay, the main way scientists used to determine "age" was by Stratigraphic Positioning. This is the determination of age by position, depth, and material consistency. MANY TIMES, this is the only method for cross comparison that was thought to be reasonable for confirmation of Radioactive decay. Scientists simply determine the depth of objects, or features near the object, or number of lava flows, or similar geologic characteristics and use the depth as a time gage. This type of comparison may not have a very high level of accuracy, but seeing things in different layers seem to show when something died. If something is lower, it is older and newer is newer. Added to this method is something called the K-T boundary, where iridium chalk was deposited from an ancient meteor that struck the Yucatan around the time the dinosaurs died. Scientists have been using this for a long time when, all of a sudden, there were trees found that were going the wrong way. **Stratigraphic Anomaly**-The next set of pictures shows some of the unfortunate trees that must have died repeatedly to be deposited perpendicular to all of the stratigraphic lines.

Some try to state the trees simply fossilized while standing upright for MILLIONS of years as the ground built up around them. [20 points on the BUNK meter!]

Distance to the Sun-If neutrinos from a single solar flare can make things look vastly older, what if the entire Earth was closer to the sun a few hundred thousand years ago? I know that sounds odd, so just keep it in the back of your mind right now. Right now, I'm going to provide you with a more logical way the Pacific Ocean was made at the beginning of the Triassic Period as our planet rotation was not stable. That is where Ice samples come in.

Ice Core Dating Correction

Let me get back to the whole Ice core testing again. Although the task is tedious, ice can be examined just like tree rings. Each summer ice changes its consistency. H_2O (16) is more concentrated in the summer while H_2O (18) is more concentrated in the winter. This gives us indication to the level of CO_2 which in turn allows us to understand something about the temperature levels. As the yearly cycle has freezing and thawing, ice consistency varies each day, seasonally, and yearly, depending on Earth axis and other critical elements. Anyway, scientists around the world started boring holes in ice. The most coring is done in Greenland and Antarctica. A sample is shown below. [This is backwards to the one I showed previously.]

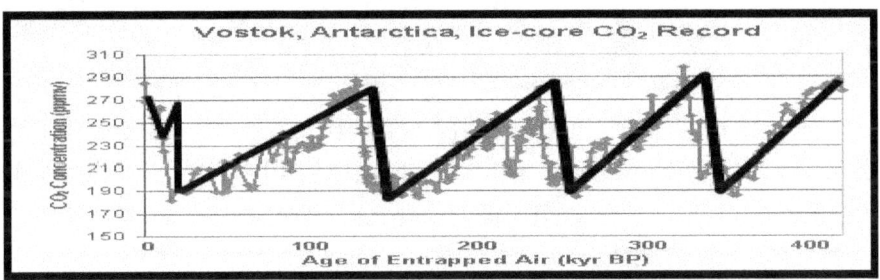

If you look closely you can see that about every 120 years there is a major change in the environment. This was found at both Antarctican and Greenland Ice cores and the dating is by seasonal changes rather than nuclear decay. Bah humb! You say! Well, what if we see confirmation?

Hawaii Hotspot Track Dating Correction

Hawaii is not a tiny group of islands, but instead is an indicator of where the Earth magma has a hotspot. As the crust moves differently than the stuff below, the hotspot relative to the crust moves and each time the hotspot burns through another piece of crust, a volcano erupts which seals off the area after a time and an island is made for a few thousand years. This travelling hotspot known as Hawaii is show next. The descriptions provided shows what was happening along the way. Because the hotspot moves perpendicular to the axis of the Earth we also know how the earth was spinning as shown by the lines in the first graphic below, but the actual timing is not described here. I placed some general times in the second graphic, but let's see if they make sense.

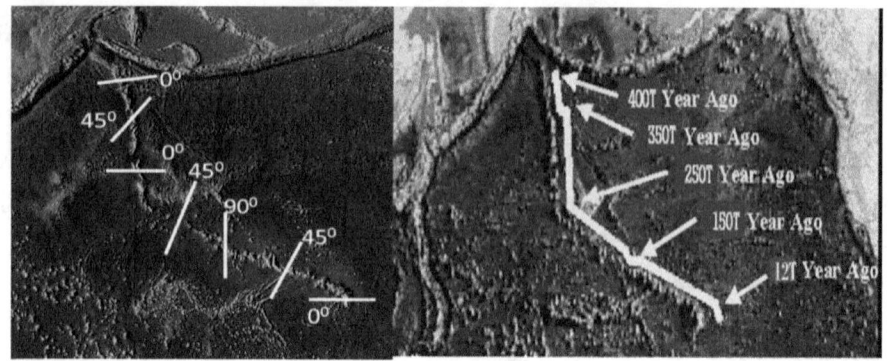

Let's compare the Earth shifts with the Ice core data. Man-oh-man; it seems they match. I think you still believe in nuclear decay so we will look farther.

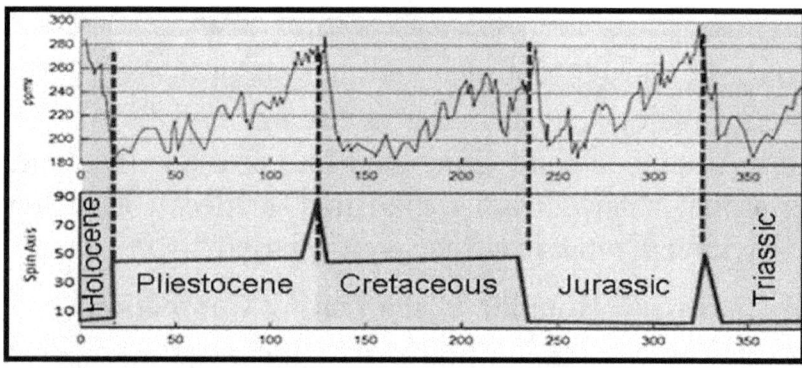

Remember the mid-Atlantic Ridge I showed earlier. While the building of the great mountains has little to do with the normal tectonic plate "drift" We can pretty accurately measure the widening ocean in various ways including measuring distances between matched magnetic landmarks on either side of a widening gap on the ocean floor. The old theory indicated at the end of the Permian, the continent Pangea began splitting apart and has been drifting ever since. In so doing, the landmasses of the Western and Eastern hemispheres separated and opened the Atlantic Ocean basin today. Plate tectonics tells us the outer hard crust of Earth consists actually of a dozen or so distinct, hard plates that drift individually on hot, deformable rock. An unequal distribution of heat within Earth moves the plates. The boundary between the plates forming the Atlantic Ocean is smack down the middle along the Mid-Atlantic Ridge, shown as the hashed line in the figure below. The ridge is where we must look to find a widening gap, which accounts for the widening ocean.

That is where we measure the rate of separation. Where the plates separate, white-hot soft mantle oozes up from great depths within the Earth to fill the gap. The molten rock cools slowly into new slivers of sea floor. This happened over and over again through the eons. That's how the Atlantic Ocean widened-by a spreading sea floor. Iron-rich rock has a peculiar property; heat it above its curie point of 580 degrees Centigrade and it loses its magnetism. When it cools the rock gets re-magnetized in the direction of the existing Earth's magnetic field. So, it's a magnet with the poles aligning with the poles of the Earth at the time of the cooling. The neat thing about this is: the magnetic field of the rock, once cooled, stays frozen in this orientation.

It becomes a record of the Earth's field at the time of its cooling.

The first graph below shows how the magnetic field has changed over time. Certainly, we cannot get an actual time, but a relative timing is very good. What if I told you this matched up exactly with the Ice Core and hotspot data?

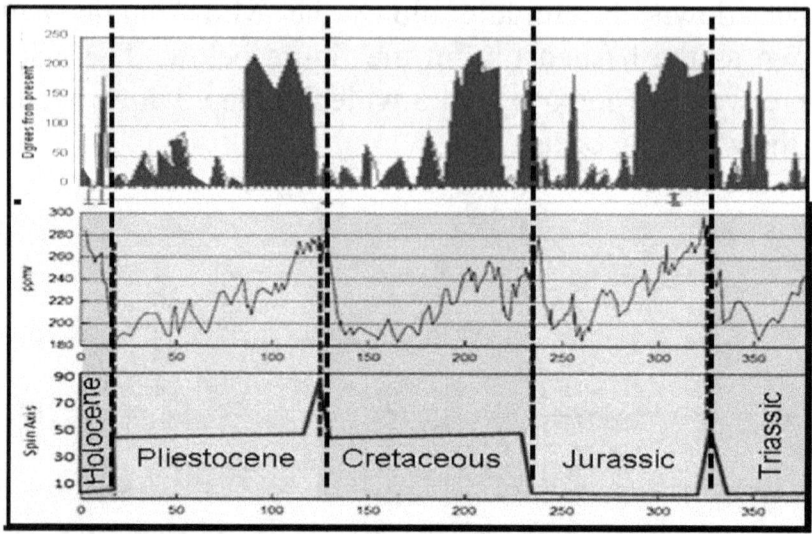

The top graph is a Archaeomagnetic graph of magnetic field changes over a set time. The cyclic nature generally matches as one would think it should with the timing obtained by the first two ways and it shows that the earth has flipped on its axis many times since Pangea began separating. Besides these three, let's look at marine life.

Marine Isotope Stage [MIS] Dating Correction

Some people may still be reluctant to give up what the schools have been preaching so very long, so I thought I would bring out one last attempt at presenting sanity. Large numbers of scientists around the globe are doing Marine Isotope Stage timing by digging in dirt. It seems looking at the levels of Oxygen 18 shows how hot or cold a point is in time while checking relative Oxygen 18 isotopes in Calcite [which just happens to be the main ingredient in seashells], one can tell just how many of the things were here during each period. Checking around the globe has given us a good map about climate and number of seashell, which correlates to number of animals in general so it is easy to see where extinction periods are. Guess where they line up? Time's up! They are an almost exact match as shown below. MIS levels are shown next above the ice core samples, the hotspot data and the magnetic field shift data Massive drops in O_{18} mean massive drops in sea shells and all other life. Notice there is no extinction period between the Tertiary and Pleistocene Ages marked by Cro-Magnon appearing. Please say you see a comparison.

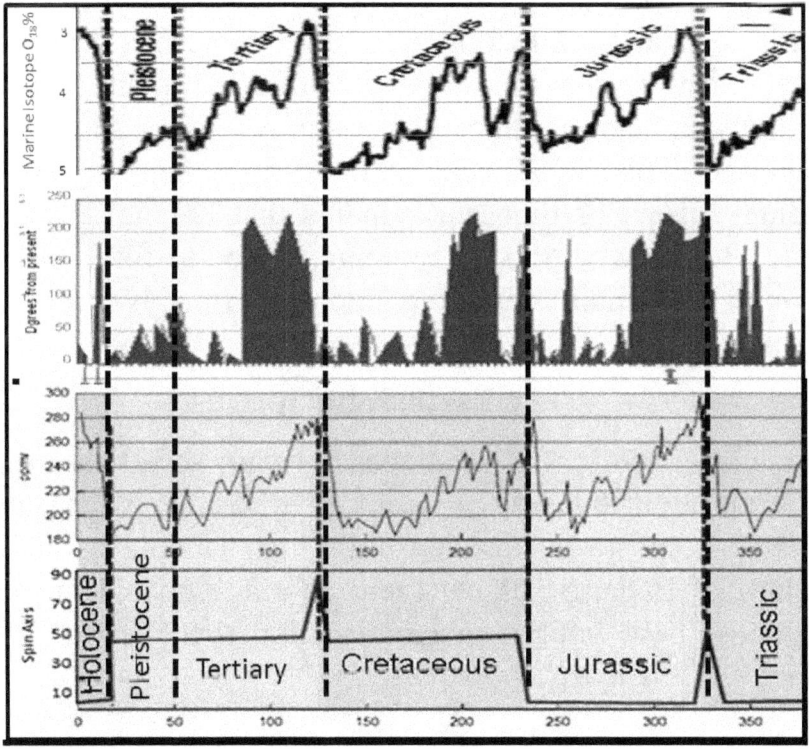

Now for the truth as we can very accurately track seasons etc. in ice core samples so we can get an absolute date to all of this. About every 120 thousand years there is a massive change in the earth's environment. This is caused by a number of things including a change in the earth's spin axis and other things we seem to have no control over. Typically, we would expect an extinction period whenever one of the abrupt environmental changes happens. I placed the "Common" names for each of the ages with extinctions and while the timings are compressed, the relative characterization of each period is similar to the unreliable nuclear decay method. We find the Pleistocene Extinction occurred 10 thousand years as the earth shifted and shifted back within a very short time. We know the massive herds of millions of Mammoths were instantly shifted to near the North Pole while they were eating flowers and millions died. The next major extinction was the Cretaceous Extinction when most of the dinosaurs died. Rather than being 65 million years ago, it appears it was only about 120 thousand years ago. This helps us understand how and why we are finding dozens of dinosaur remains that are not fossilized today. Before we go to a new topic, I need to fix this stupid theory about India slamming into the Himalayas to make mountains.

The Earth Split Open

This crazy theory was developed by vain scientists to hide the truth about our Earth. Again, plate tectonics smashed a massive island to make, the 6-mile high Himalayas and the country of India. Strangely India, is made up of almost all magma so how would a magma island move. [See the island smash "lie" below.]

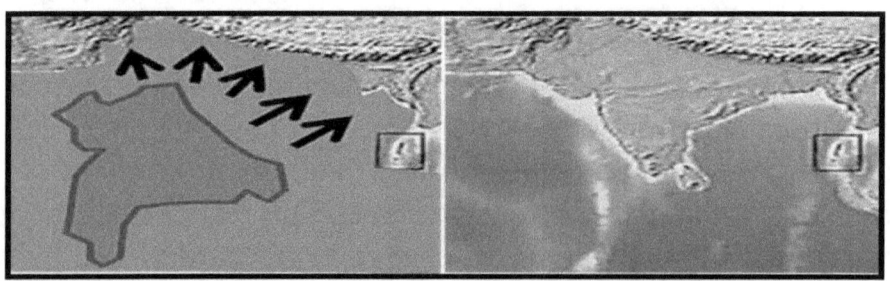

What happened has happened at least 5 other times as massive meteors crash into our planet. One crashed at the end of the Cretaceous near Yucatan with such force, the other side of the Earth split open and millions of cubic miles of magma and iridium dust escaped. The magma made the Indian area known as the Deccan Flats which takes up about ½ the country. I guess false historians don't want you to think the Earth could be split open.

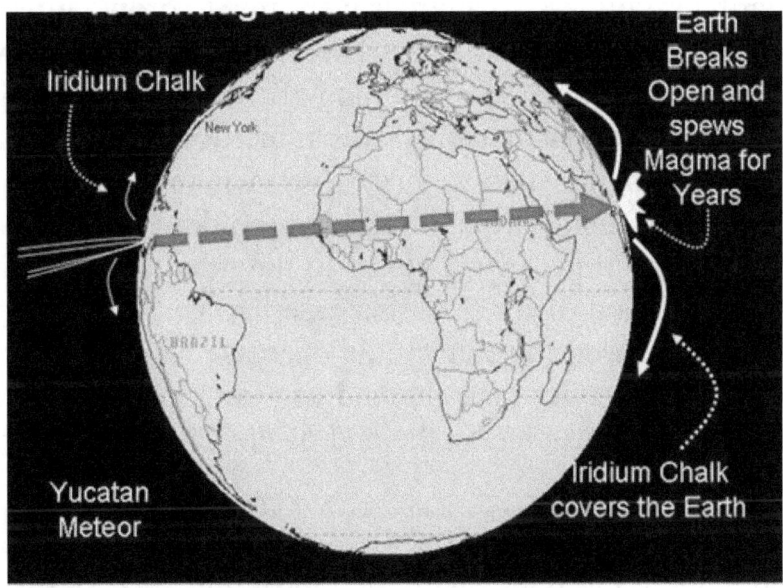

How Much Dirt Did We Lose?

Our earth now has a volume of 288 billion km³. Most of that is called the Mantle with the core also being 38 billion km³ and the crust being about 45 billion km³ [the outside part where people live and plants grow]. Before the event we had an additional 45 billion. Of that part of our planet that is missing, a portion collected to form Lunar, our largest moon; approximately 20 billion km³. The remaining 15 billion went into out space. No matter how you look at it that is a lot of rock, dirt, and animals. Even today, the massive hole is so scary just knowing that there is almost no crust below ½ of our planet. It's all gone! Teachers teach a lie by not providing the details.

What Other Ramifications

Just like any other spinning sphere, as our planet got smaller, its rotational speed got faster which would have caused some of our atmosphere to begin leaving Earth just like the rotating Saturn's atmosphere is leaving it today. This faster spin made the Earth gravity less so animals were free to grow larger and larger during the Mesozoic. The weather patterns were completely disrupted, so everything about the earth changed. We can believe that not only was rain substantial, but also volcanic action must have been everywhere. Just like any other mass spinning around an object holding it by gravity, the planet would be driven closer to the gravitational source over some period of time. In this case it was the sun. As our planet came closer, the spin would decrease which eventually stabilized the gravity near the end of the Cretaceous Period. Everything was heavier by the end of the Cretaceous period, or at least

I can't tell you a good reason why teachers are not telling our children about the issues with the old timing and not rushing to this tested, verified, and correlated timing model so we can better understand our world except that books will have to be rewritten and vain scientists will have to tell the truth for a change. Speaking of telling the truth, the earth rotational axis is unstable as we looked at with the Archeo-magnetic details of the mid-Atlantic Ridge.

Another thing that is not talked about is how our earth rotation is not constant. When we go faster, everything is lighter. When our planet rotates more slowly, people and animals get heavy and some die.

Fast Earth Theory

Current Theory is the Earth rotation is and always has been the same. Therefore; dinosaurs were too heavy to move.

This one is pretty much a Preston exclusive, but I think it should really be investigated. You have been taught some pretty stupid things concerning dinosaurs, so we need to straighten them out. No! No! No! flying dinosaurs were not forced to walk around and climb to the top of a cliff so that it could glide to the ground hoping to catch some food on its way down. The Tyrannosaurus Rex was not so weighted down and clumsy that it could not kill live prey and the diplodocus was able to lift it head all the way out at the end of its long neck and its heart COULD pump blood all through its body. The picture given to us by so called scientists is an impossible world filled with impossibly weak monsters that somehow lived even though they could not reasonably survive.

In my infinite wisdom, I determined that if a description is shown to be almost impossible, there is a good chance that it IS impossible. So, it is with the world of the dinosaurs. Their life was substantially different than the description above. After many years something killed them, but that also was not what you have been told. First you were told a massive cloud of Iridium dust encased the Earth and killed the dinosaurs, but the dinosaur bones were underneath the Iridium dust [called the KT layer which somehow means Iridium] The dinosaurs died before the Massive meteor strike we discussed earlier. Then you were told that the tiny animals killed the big ones or the huge dinosaurs starved and the tiny animals stayed well fed. The general rule of thumb is that larger animals survive better than small ones. Not the other way around. It was no different during the time of the dinosaurs and for almost the entire Mesozoic Era the mighty dinosaurs NEVER lost their place as rulers. As far as a climate change killing the cold-blooded lizards and letting the warm-blooded creatures live free, there is substantial evidence that many of the dinosaurs were warm blooded. Tiny egg eating dinosaurs might have been a nuisance, but a couple of bites and they would have been history. No! it was something else besides dust, starvation, tiny animals, and weather that destroyed the dinosaurs.

There is only one real explanation for the strange event. If these huge animals survived during the Mesozoic, they could not have been as heavy as their masses would dictate. That can only mean that the earth was spinning faster so that the creatures on it would not be as heavy. I know it sounds simple, but some still try to ignore reasonableness and I have not found even a suggestion of this commonsense reason in ANY historical textbooks.

The Martian close encounter and sucking out the continent of Prestonia would have reduced the mass of the Earth and it would have certainly spun faster. Evidence is shown as animals and humans got really big. There was a bad side of this in that our atmosphere would have slowly escaped into the solar system just like the atmosphere of Saturn is today. Tests on atmosphere captured in amber confirm the high concentration of Oxygen at that time.

At the end of the Cretaceous, when the meteor hit or possibly an earlier meteor hit caused, the Earth rotation slowed down again to where it is today. Today, we are not losing our atmosphere and physiological studies of dinosaurs show they could not have run, raised their heads or flown as they must have been able to do before the Cretaceous Extinction [120 thousand years ago].

Dinosaur Size Evidence-During the Jurassic, Triassic, and Cretaceous periods, the largest animals that ever walked on land were in abundance. The bone structures of these massive monsters don't exactly fit the animal structures that would be required today. If there is an anomaly it probably means that we haven't uncovered the truth.

Diplodocus Neck Evidence-One such anomaly is shown with the Diplodocus. Its neck is too long. If the creature had tried to stretch out its neck in front of its body, the head would have come crashing to the ground as there is not enough muscle to pull it into the air unless its neck was lighter.

Tyrannosaurus Rex Arm Evidence- The Tyrannosaurus Rex structure is such that it would not have been able to run in our atmosphere. If it couldn't run it would have died. It is said that the arms are so short that if the T-rex ever fell, he would not be able to guard his fall and his huge head would be damaged in the fall. Even getting up would have been a horrible thing if his head was still working. Many have relegated the T-Rex to a scavenger like a vulture. Let's face it if T-Rex could not run after food it would not have survived nor would it have needed those monster teeth. The T-Rex wasn't worried about his head because he was not very heavy.

Bird-like Reptile Evidence- Many of the Bird-like reptiles would have surely died as they could not fly in our current atmosphere. To make this element more comical, as I said scientists invented many theories that suggest that the winged reptiles would have had to climb up to the top of a cliff and take off into the air. If they ever hit the ground they would be doomed as they could not regain flight. I know it sounds stupid, but that is the belief of many today. Hopefully you can appreciate the absurdity of such a theory, and begin believing in a faster spinning earth.

Dinosaur Heart Evidence-Still another problem is the dinosaur heart. It has been determined, by many, that the hearts on these massive beasts would not have supported the blood flows needed to sustain body function. Many have surmised that the dinosaur's blood flow problem could have been rectified by increased oxygen percentage in the air. The blood flow could be greatly reduced if more oxygen was present in each breath, but that does not account for many of the physical characteristic anomalies, and I'm not sure it would even satisfy the blood flow problem without other elements like lighter animals and then you have the question of why the oxygen was here and not here now. Think I can bring some sanity into this whole crazy mess.

Moon Evidence- If we make the assumption that the specific gravity was less during these ancient times, then there must have been a cause and we don't have to search long. Generally speaking, less gravity means smaller size, less density, or faster rotation. I don't believe that the earth got larger during the Mesozoic, so the other two seem more likely. Concerning density, we have one very nice piece of evidence—the Moon. Below are three of the elements that we believe we know concerning the Moon and its relation to the earth.

The Earth has a large iron core, but the moon does not. The Moon mass is only about 60% of the earth's. This is because

Earth's iron had already drained into the core by the time the moon was yanked away.

The moon has exactly the same oxygen isotope composition as the Earth, whereas Mars rocks and meteorites from other parts of the solar system have different oxygen isotope compositions. This shows that the moon, most likely, formed from material in Earth's neighborhood.

The moon orbit and Earth rotation are synchronized, suggesting that they both came from a common source.

I have presented an argument that the moon came from the earth, so let's just say that the Pacific Ocean opened up and spewed out the moon. The moon density level tells us that only the lighter portions of the earth were expelled, leaving the much heavier mantle intact. Therefore; the specific gravity associated with the earth was much higher after the expulsion than before it and Dinosaurs would have been lighter before the Moon was made than equal sized animals from after this extreme event that reduced the earth's density. Because today, the Moon is only 40% as dense and the Earth, we can make a reasonable guess that the density of the earth was perhaps 10 to 15 percent less than it is today before that section was torn away. Therefore; the dinosaurs would weigh as much as 10 to 15% less than they would today because of this factor alone, but that would not be enough difference. There must be another factor to be considered.

Saturn Evidence--The lower density wasn't the only thing that lightened the dinosaurs. We also have a pretty good example of what the earth might have been like with respect to rotation if we look at a picture of Saturn, shown in the following graphic. Saturn is wide at its equator because it rotates quickly. The Earth may have looked this way before the great slowdown 120 thousand years ago.

Saturn is spinning fast enough so that particles do not have to gain substantial amounts of energy to attain what is sometimes called escape velocity. Saturn's day is less than ½ of a "current" earth day and the whole planet is trying to escape. The bulge at the equator shows two things. It shows that components of the atmosphere are drifting away from the planet and it shows that gravity is lower. Remember the "atmosphere drifting away concept for the topic below". What we find in our solar system is that, even though Saturn and Uranus are substantially larger than the earth, things on those planets are lighter than they are on earth. This is due to the high spin rate. On Saturn or Venus, animals would weigh only 90% of what they weigh on the earth and on Uranus they would only weigh 80% of their "Earth weight".

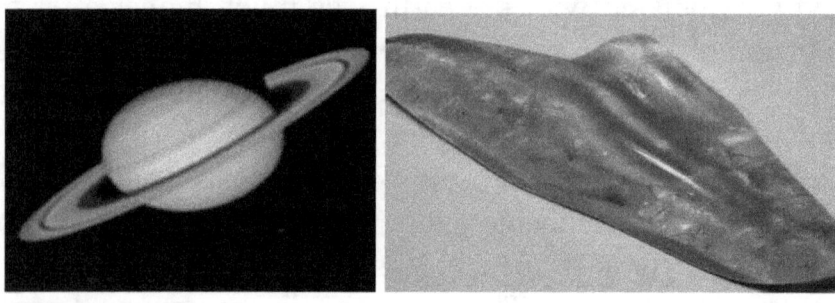

Oxygen Evidence-One reason we can believe the earth was spinning fast is the thicker air that has been lost over time. We can also surmise, from evidence described below that before the earth slowed down, it probably had a thicker atmosphere rich in oxygen. The most logical way for the oxygen level to have slowly been reduced over time would be that our planet rotation was significantly faster than its present rate and similar to that of Saturn. Here is what researchers have found concerning the oxygen levels during the dinosaur days. Experiments have been done to show that the oxygen content of the "Dinosaur's Air" was higher than our current air. This was easily accomplished by simply testing the oxygen content of captured "air" in ancient resins like bubbles in amber as shown in the preceding graphic. The tests showed a substantial increase in the oxygen content when compared with "Modern Air". The problem with this experimental data is that the researchers only used the data to try to prove that a "water canopy" was around the earth before Noah's worldwide flood. Because the canopy theory had other problems, the issue was essentially disregarded. Because the air was thicker, the earth was most likely spinning faster and eventually lost the "extra" oxygen before it slowed down.

Flying Birds with Solid Bones Evidence-- The flying lizards weren't the only creatures with anomaly. The Archaeopteryx was a solid boned bird. It could not fly with solid bones, so it must have walked around and just had feathers for looks. Remember the flying reptiles that couldn't fly? If the earth was spinning faster it would be easier to fly and the flying solid bone birds wouldn't die. Imagine that! The Archaeopteryx must have flown and its wings were not useless as you have been told.

Giant People Evidence- The Bible tells us that in a distant time before Adam there lived Giants that ruled the earth. The same story is repeated over and over again around the ancient world. If we look around we find that these giants must have lived during the time of the dinosaurs. Footprints have been found integrated with dinosaur footprints, they have been found in rock and they have one thing in common. They are all huge. Even the 8-foot people we call giant today have a terrible time. They are simply too big for our environment. During the Mesozoic, giants were not limited. If these people were as huge as their footprints indicated, they could not have lived on this earth the way it is today. By the way, I'm talking about evidence of humans over 20 feet tall have been found. Just think about a guy over 2 stories high trying to get his lungs and heart to work.

Earth Stability Hypothesis

Textbooks claim the Earth rotational axis is solid and the truth is almost completely hidden. I have no idea why?

From the Archaeomagnetic chart in the preceding graphic we can see that the earth has actually shifted on its axis about 170 times over these shifts are the past 350 thousand years. Sometimes these shifts are temporary and sometimes they almost flip the earth magnetic field or rotating axis. The events causing major shifts in rotation are also recorded in the Hot-spot trail data. It seems major axis shifts are a major reason for extinction. From this timing component we can see that a major change occurred about 11 thousand years ago followed by a reversal about a thousand years later scientist call this time the Young Dryas. The end of the Dryas marks the end of the Pleistocene and the worldwide flood extinction. That being said, the earth did not shift 180 degrees; instead, it shifted about 30 degrees and a thousand years or so later the magnetic field shifted back without the Earth axis going back as massive tidal waves swept the entire planet and very few survived. Noah was one.

Just before this destruction is a time we find something called the Young Dryas. While Dyra is a flower, the time was no sweet-smelling loveliness. Instead it was a wild time and it took a long time; on the order of a thousand years.

Imagine a 1000-year war! It was the end of the last Ice Age. Temperatures increased steadily for the next thousand years, but this is where it starts getting interesting. Levels of copper, tin and lead show marked increases and there is this abrupt ending that looks really weird which has a huge drop in temperature as measured in Greenland and Trinadad, snow accumulation in Greenland, and O^{18} density levels in Antarctica. Something major happened even before the Pleistocene Extinction. From much data we find a massive war was raging for years before the Extinction, but most forgot to report on this important information.

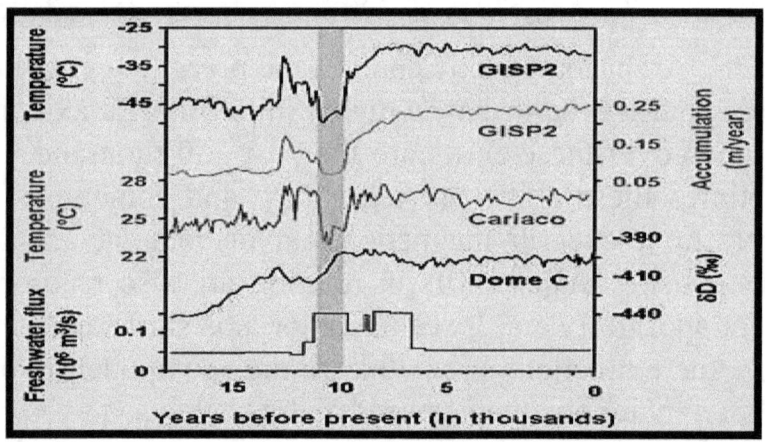

Death was everywhere and then we find something very scary. Uranium concentrations in coral jump by almost 300%. Also, we find marked increases in nanodiamonds, magnetic spherules [tiny balls], and carbon spherules at the end of the War with a major increase in charcoal around the middle showing fire unbearable heat and nanodiamonds indicating nuclear explosions confirmed by the uranium concentrations found. ----- This occurred during a fairly brief time in our history as shown next as depth can be interpreted as time and the Nuclear fallout occurred during the Young Dryas.

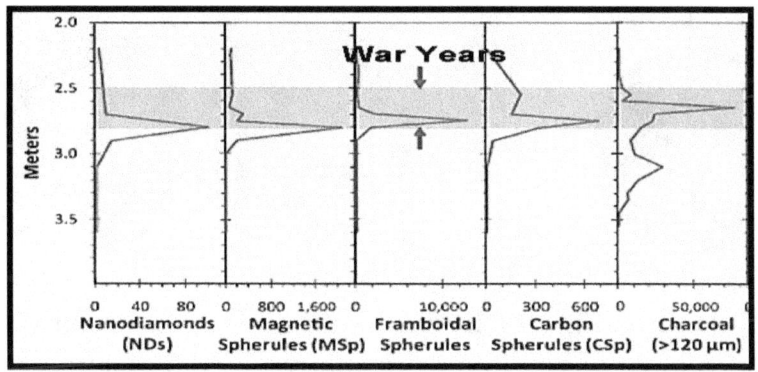

Miles of fused desert sands in Libya and Egypt may also show a massive high temperature "explosion" without a meteor. All the evidence seems to support the existence of an ancient nuclear war having taken place at this time and felt around the world.

Soon after the war, something very strange happened, Venus was destroyed and over ½ million meteors from Venus, or its no destroyed moon, pelted the East Cost of the United States just before the extinction and worldwide flooding.

Hawaiian hot spot details- As I discussed earlier, the Hawaiian hotspot not only shows changes consistent with the changes noted in the Ice Core but also the angular changes in the tracks can be reconciled as part of the rea son for extinctions. The axis shift 10 to 12 thousand years ago shows the 30-degree shift did occur, however, there seems to be no shift back as the first graph suggests. The images below show the major axis shift and the image to the right shows to possible location of the poles during the Pleistocene.

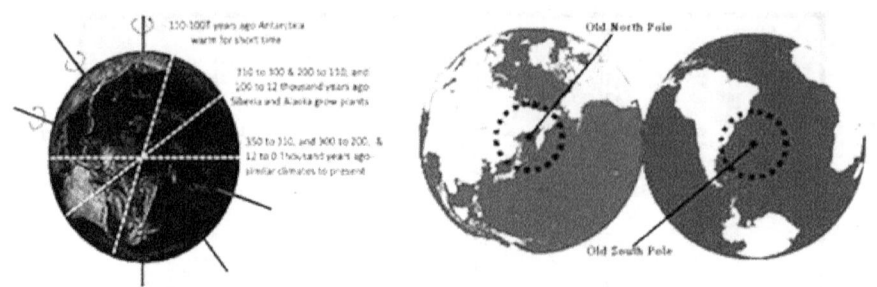

Carolina Bays line of bombardment- The east coast of the United States was pelted with many objects. There are still an estimated **500 thousand** meteorite craters called "Carolina Bays", which mark this incredible event in history. Some have a diameter of over 14 miles. Just think about how afraid the people of that time were as they essentially saw the sky fall all around them. The graphic below shows the major areas where these objects have been found in the United States. These generally date around the same time when just before its axis shift by 30-degrees to where it is today. This and Pleistocene materials found in the crater structure show it happened just before the extinction. A collage of a couple hundred craters is shown below. Evidence in Australia shows a "straight-line" distribution pattern on the other side of the world that is that is consistent with a bombardment along the equatorial boundary. If we consider the impact density line as the "Old" equator, a shift of about 30 degrees in the rotational axis has occurred since the bombardment as shown below. The globe has been separated at this ancient equator. Note that along the equatorial path there was not much land. Also note that eastern Alaska and Siberia are well away from the Arctic Circle, which allowed huge herds of Mammoths to dine on flowers in those areas just before the shift. The shift froze them solid.

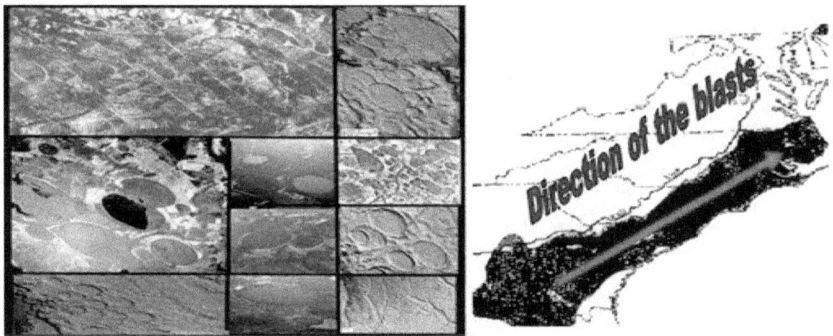

There is something else the Carolina Bay craters tell us. There was a disturbance that could have had enough power to shift the Earth's axis during this time.

Frozen Mammoths- Not only are massive herds of Mammoths found in Siberia where they should not have been, but also some still have flowers in their mouths and there are estimated to be about a million in the massive herd showing Siberia once was a vast meadow meaning the earth axis was significantly different.

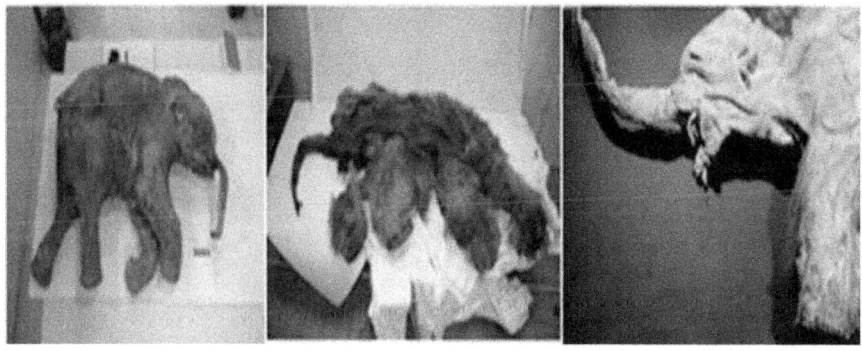

We can almost see the destruction as ½ million meteor hit and turn the forests to fire as noxious plumes rise and the earth shift sets of floods, unbelievable tidal-waves that swept the mountain tops, and destruction everywhere. Soon the entire world would flood as the polar ice caps reform but many try not to even believe that there was a worldwide flood at all as the Pleistocene extinction ended. This would give credence to the Biblical history over their own vain of how have made up their mind concerning the past.

The shift in the rotational axis melted the poles and the new poles quickly began to refreeze, but in the meantime massive flooding was only the beginning and tidal waves over a mile high smashed across what was left of the dry land. Even the mountain tops were hit and just about everyone and everything died 10 thousand years ago.

Well after the Pleistocene extinction, men regained civilization but then 5 thousand years ago a war destroyed infrastructure and a massive mutation of human DNA caused man' advancements to be lost for a time. That is not to say there weren't maps that survived and map makers traced the maps showing the new world. Soon a Greek mathematician not only publicly described the Earth as a sphere, he measured the thing and many adventurous seamen set sail to find the Americas. Christopher Columbus was not one of these guys. To make his story sound more real, history book bring out this bogus idea that the people thought the Earth was flat and sailors thought they would fall off if they went to sea.

Flat Earth Theory

This theory simply states that the Earth is flat. Supposedly, Columbus was trying to prove this to be false when he discovered the Americas. Both, of course are bogus.

Greek Mathematics

In 240BC, a man named Eratosthenes was the head librarian at Alexandria. He decided that navigation to other places would be easier if navigators knew the distance around the earth. He rediscovered the circumference of the earth to be 25 thousand miles and also determined the earth's axis tilt to be 23 degrees before publishing the details in the book entitled "Geography". This work was the standard for many years and only very recently has our data been updated to have the earth's circumference at about 24,900 miles. All navigators used this data and knew it to be correct for many years.

Going Flat

Everything was going OK until this African guy came along, named Lactantius. He developed his flat Earth theory around 290AD. That theory was wrong, but it didn't stop people believing it because Lactantius was a great religious leader and he told everyone that it was stated in the Bible. Of course, it wasn't; but Lactantius wanted it to say that so much that he convinced himself and he messed up some. Don't worry about him so much as his theories were ignored by the sailors of the day because they needed Eratosthenes' calculations to find things. Others saw the light and his views were soon considered heresy by some of the Church Fathers and his work was ignored until the Renaissance.

Renaissance Revival

This is a perfect example for this book. While trying to translate all things Latin, Lactantius' "Flat Earth Theory" was translated. Because it had been in Latin the renaissance people believed it instantly, except for the sailors who needed Eratosthenes' calculations to find things. That brings us to Columbus.

1492AD Columbus Theory

Columbus is known for a lot of things, but are they the right things? What we really know about him is that he was a fantastic liar, introduced young, Caribbean, slave girls as a money crop. He was also responsible for the new world's first "lost colony"; and more than likely, introduced syphilis to Europe. To top it off he was not a very good captain or navigator by all accounts, but he certainly knew about Eratosthenes.

Columbus The Liar-Back when Christopher Columbus came along, no one wanted to fight the Moslems who were taking over the Mediterranean coastline. Therefore, people started to gain land other ways and the "new age of exploration" was born. No explorer is better known than Christopher Columbus, but there were several sailors trying to do the same thing as old Chris during this same time. Columbus was just the biggest liar. By manipulating known and calculated distances, he convinced Queen Isabella of Spain that the distance from Europe to "the Indies" was only 2500 miles and that it could easily be sailed to in less than two months. He knew it would take much, much longer, but he wanted the Queen's money. He also carried the lie much further by not letting his crew know what they were in for. He modified the ships books to keep his crew from knowing anything, so their trip was a complete horror, but it is apparent now that he probably knew all along how far he had to travel. He was trying to get to the huge deposits of copper that were being brought back from the Americas along a northern route at that time and he needed to get there before others found out about. I know you were told he was trying to get spices from China, but that simply could not have been right.

Isabella The Kind Hearted-Isabella had already had all types of fun by initiating the Spanish Inquisition in 1478 and by 1492, she had finally gotten rid of the 100 thousand pesky Jews that kept trying to stay in her country, so the Moslems took them in. [That doesn't happen too often today.] That same year, Christopher wanted to go find the land on the other side of the Atlantic and she had done everything else, so this was a natural. Other greedy sailors tried to gain their support in Spain, but had calculated distances of almost 10 thousand miles and the investors were not certain that the trip could be successful. Lying was the only solution.

No China Trip Theory-Christopher's trip had nothing to do finding out if the Earth was round nor did it have anything to do with going to the Orient. You may have been told these things, but don't have anything to do with the evidence. No "novice" navigator of the day would think that China could be so close and Christopher was no dummy. OK he wasn't the sharpest tack, but he had been well educated as a navigator. Just by measuring the time it would take a ship to disappear over the horizon would have allowed most, including Christopher, to know the approximate dimensions of the Earth and of course, Eratosthenes's very accurate measurements would have shown that trip to be over 10 thousand miles and impossible, except that most Northerners knew about the copper rich Americas and that is where he was going.

American Map Theory-Not only had almost every navigator determined what the approximate diameter of the Earth really was, but also, they had maps, by all accounts. Several maps of the world were still extant from ancient times, which showed the mid-ocean landmass that Christopher was after. We do need to know is that the distance from Europe to the "New World" is approximately 10 thousand miles and Christopher must have known it. Below is a map made in 1395 by the Zeno brothers of Venice, Italy. They had traveled to the Americas along a similar path that the Norse had taken. This map would have been generally available to the navigators of the day

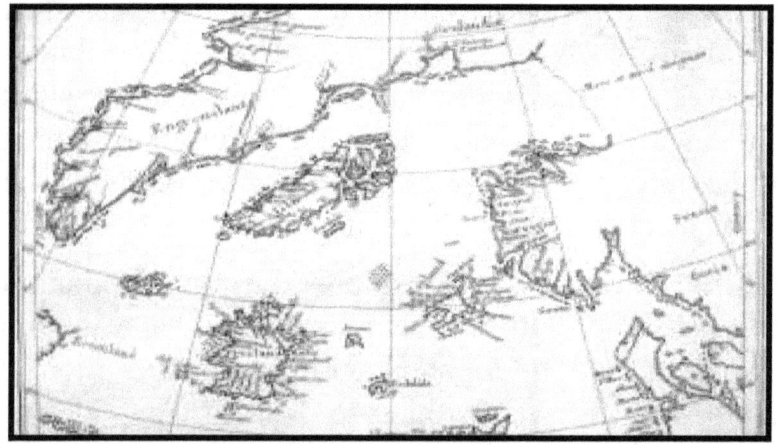

Columbus the Navigator-Speaking of lying; Chis's real name was Domenico Colombo, but the Christopher sort of stuck as a parallel to "Christ", I suppose, and he turned out to not be such a great sailor. In 4 voyages, he lost 9 ships. That was a terrible record. In fact, when he died, his funeral was not a lavish affair as other great navigators and explorers of the day. He was only a minor player by all accounts, but we praise him in every history book.

Domenico Colombo alias Christopher Columbus

No one knows what he looked like, but the drawing above is the best guess we have from a number of descriptions.

Columbus the Slaver Theory- So what should Columbus be known for besides lying? In the early 16th century, Columbus tried to get to India, landed in America, and brought back his greatest discovery. He found that the young girl slaves didn't die as easily on the trip back to Europe as male slaves and they also brought the highest price. When he got back from his first trip, everyone was anxious for him to return and bring back more of this wonderful commodity. So, he went back and loaded up with more female prizes.

Columbus Found Syphilis Theory-At the same time that the slave girls were introduced, syphilis, all of a sudden, came into existence in Europe and nobody knew where it came from. Evidently, there was a substantial difference in the people of the two hemispheres and some diseases were more severe to the west and some were more severe to the Eastern populations of Europe. Syphilis was a terrible price for having American slave girls.

There are many things to say Columbus did, but finding America simply was not one of them. The Phoenicians found it accidentally around 600BC and many others followed. By 1492, many maps and trade routes had been established and used for years. Columbus simply conned Isabella. Then his sailing skills were so bad, he almost killed himself in the process of reopening trade to the new world.

Columbus was a rat, but today we have similar rats trying to cash in on fear. Knowing their ideas are bogus, they pushed something called Global Warming.

Global Warming Hypothesis

This theory states that the world is on its way to becoming the massive fireball that Venus has become. A corollary indicates that man has some special destructive quality that is making the Earth do this. Corollary #2. Burning coal and cows farting are the main causes for our eminent destruction.

This theory goes from stupid to criminal as details were trumped up by a wide assortment of so called scientists to allow for increased power, prestige, and money. The whole scandal is starting to come out with the embarrassing release of emails from the University of East Anglia Climate Research Unit which showed their part in the misconceptions hiding research results that were detrimental to the claims of man-made global warming hypothesis. Then came scientific determination of destruction of glaciers. This was also bogus. It was report that the Himalayan glaciers could all disappear by 2035. Of course, it is impossible, but that didn't stop the contrived disaster. The IPCC took its best shot. According to the "Telegraph" we read the following.

...not content with having lied to us about shrinking glaciers, increasing hurricanes, and rising sea levels, the IPCC's latest assessment report also told us a complete load of porkies about the danger posed by climate change to the Amazon rainforest.

Ice Evidence-That brings us to the ice core samples we have been taking for years. They give us the history of our planet's temperature. Rather than taking the word of the self-proclaimed inventor of the internet let's see what the ice tells us. Many use a very erratic Greenland graph alone to show global warming or a very small portion of the Antarctica graph to show a similar event. If I was trying to find a heating cycle in ice core samples over a 120-thousand-year period below left I could not find one. One the right I have a sample from 6 sites on a single graph and they show massive reduction in temperature since 12-thousand-years-ago. As I showed in previous Ice core samples this 100-thousand-year cycle has been going one for hundreds of thousands of years and they were not affected by dinosaurs or civilization.

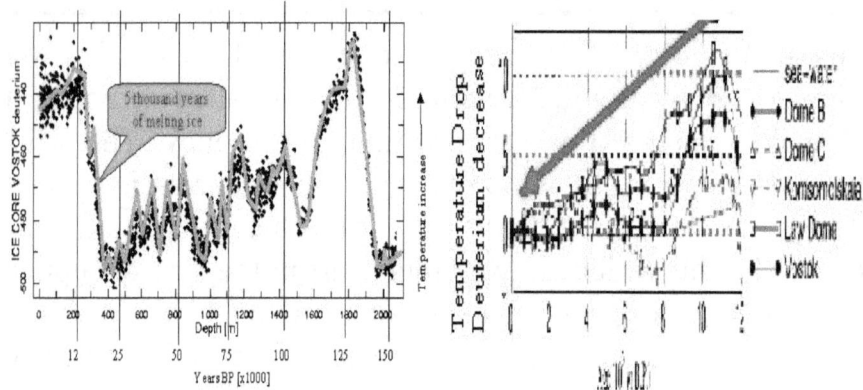

Please don't let people tell you what "arbitrary information" means without showing you the data and if they do show you data, please go check it yourself as many will go to extremes to push an idea, especially one that has such a massive price-tag and global warming.

The Earth is getting colder!!!!!!!! Except during short periods of high sunspot action.

If you would like a refresher graph, the following is similar to previous ones showing the cyclic temperature of the Earth over the past 450 thousand years or so.

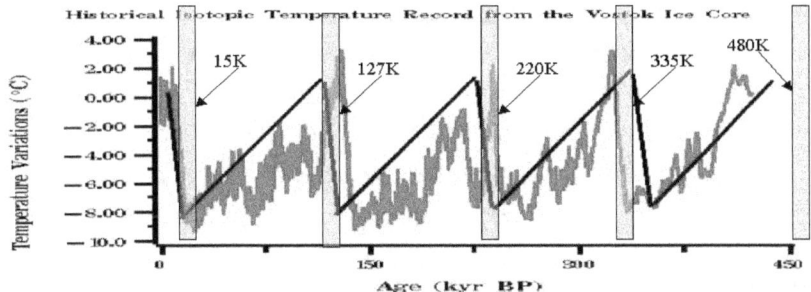

We have all been hearing it. Global warming, Global Warming! Quick get rid of you underarm spray as it might let hydrocarbons into the air! Even with 50% of our Electrical energy coming from coal, we need to eliminate the use of burning coal or all our lands will dry up! Now build electric cars to eliminate the destructive use of gasoline. Wait a minute! I'm back to coal again making electricity. All Greenhouse gases except for the only one that could possibly cause a problem became the villains of the 21st century as they pushed us closer to destruction.

Worst Greenhouse Gas

H_2O gas was ignored even when everyone knew it causes. so much temperature change because no one could tax it or make money from demonizing water vapor. You also hear almost all scientists agree that human caused global warming will destroy the earth when 30 thousand scientists signed a petition against the horrendous statement.

Who Are Self Proclaimed Climatologists?

While many indicate they are climatologists, the word is nebulous and easily mischaracterized. There are climatologist claims from individuals with degrees in mathematics, atmospheric sciences, meteorology, oceanography, physics, chemistry, geography, geology, astrophysics, statistics, astronomy, engineering, earth sciences, environmental sciences, and more specific fields within those already mentioned.

> *Scientists like Al Gore told us <u>the Earth is dying because of man</u> and we gave him the Nobel Peace Prize even though he only brought loss of peace, an abundance of fear, and made a $billion by buying quasi-green industries and proclaiming the end of the world.*

He went to accept the tribute in his massive, private, fossil fuel guzzling jet and praised how we were turning a corner with horribly inefficient windmills, minimally effective solar panels, making plants artificially grow faster with DNA modification to reduce the need for fertilizers, and on and on. This is such an important issue; we must make a determination of truth. Unfortunately, there are many levels of truth in a world of what we can call vain truth. Many time vain scientists HELP their vain truth look like reality.

Climate is Changing

We know something is happening, but what it is and how it will affect us needs to be understood without shouting unfounded claims to support some pet project or money scheme. While our temperature has not been affected, greenhouse gas concentrations SEEM to be steadily going up slightly over the past 50 years as shown next. With great conviction some tell you reasons, but no one actually knows, for sure why it is happening and what effect, if any, this increase will have. It should be noted that while these increases have been steady, the gases identified here make a **tiny** portion of our atmosphere---I mean less than a tiny portion.

Antarctica Ice Loss Lie-Even the government paid group that is supposed to protect us where climate is involved got into the changing the data to force fear. The first graph shows NOAA data to push fear, while they also published the second one and never claimed it meant anything. Of course, the first one has been manipulated by showing a very small snippet and selecting a single point on Antarctica that had this temporary reduction.

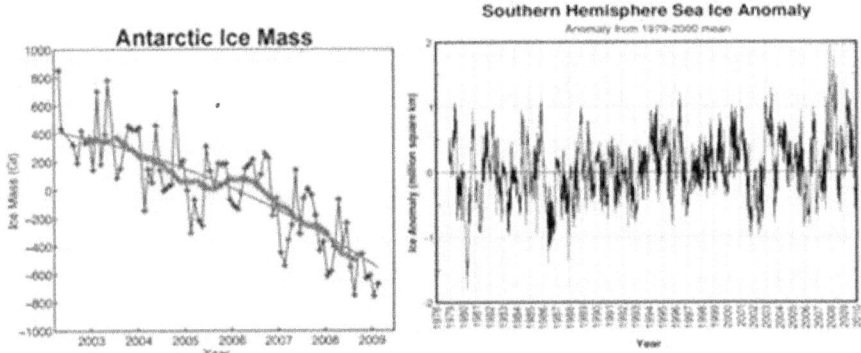

Sea Surface Temperature Lie-Here is another alarming bit of data passed on by those with ulterior motives. The first shows NOAA **calculated** destruction of our Oceans as the seas quickly overheat and the water levels rise. Let's look at a second set of **actual** data from satellites. The Sea surface temperature as measured by NASA satellite which shows a general downward slope, but that isn't all. The average from 1998 till now has been drifting downwards and since 2009 the temperature has been declining continuously.

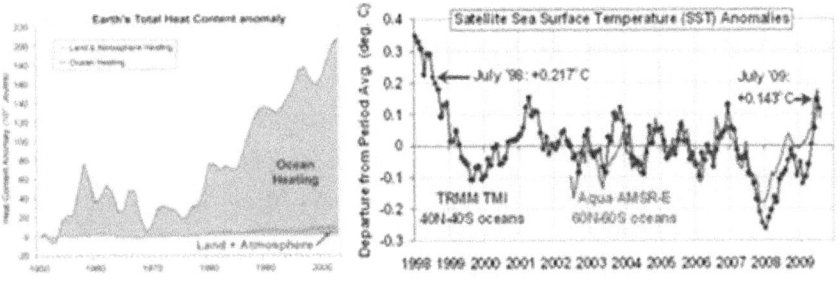

CO₂ levels in the Ice Core Lie- The next one is really sick so go through up first before reading the data. Our great friends at NOAA and IPCC took the thousands of years of Ice core CO2 levels and added the atmospheric CO2 levels from Hawaii to make it look like the re was death from people burning coal. The problem is that airborne CO2 doesn't fall to the ground so it could not be captured. First look at the scare graph and then I have other data. The second graph shows where the Ice Core ended and how the airborne data was spliced.

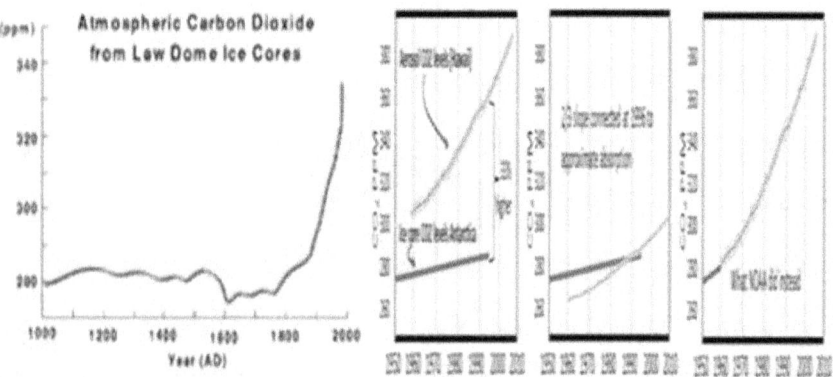

IPCC Data Tell Us The Treachery-Luckily for you, you don't need to take my word for anything. The IPCC told on themselves in a paper presented in 2009 that tried to confirm massive increases in CO2, but what they ended up showing was TREACHERY. Here are some snippets. I have not put them together just to make it look bad, but I didn't think you would want to go through all the mess so I boiled it down a little. The IPCC acknowledged CO₂ has something called a short residence time, stating:

*"The turnover time of CO₂ in the atmosphere is about 4 years. This means that on average it takes only a few years before a CO₂ molecule in the atmosphere **is taken up by plants or dissolved in the ocean**.*

As you read through this remember the Hawaiian Aerosol CO_2 detection is the thing that is causing the entire ruckus. The data taken included substantial amounts of CO_2 that would be absorbed into plants and never get into the Antarctic Ice.

"The CO_2 response function used in this report is based on the revised version of the Bern carbon-cycle model used in Chapter 10 of this report. --About 50% of a CO_2 increase will be removed from the atmosphere within 30 years and a further 30% will be removed within a few centuries. The remaining 20% may stay in the atmosphere for many thousands of years".

CO_2 Absorption lie- Besides all that calculation of how long CO_2 stays in the air, it can't absorb in the air easily. Those suggesting CO_2 is a major greenhouse gas, seemed to have never even looked at an atmospheric absorption map as shown next. Water is by far [Over 99% of absorbed energy] the most absorbed large molecule gas that can cause horror or blessing. It is very difficult to absorb CO_2 into air. Even in areas where CO_2 can absorb, water is already so absorbed that it still has an inability to be introduced in our atmosphere. Here is what we can suppose about the large increase in CO_2. There has been a tiny reduction in water vapor which shows up as a much higher CO_2 percentage. No matter what, we must understand that water vapor controls our atmospheric temperature----not CO_2. CO_2 has only 3 small wavelengths of absorption while water is absorbed in huge amounts. The second darken area in the graph is the only absorption wavelength of CO_2 while the top curve is the total amount of energy in the atmosphere.

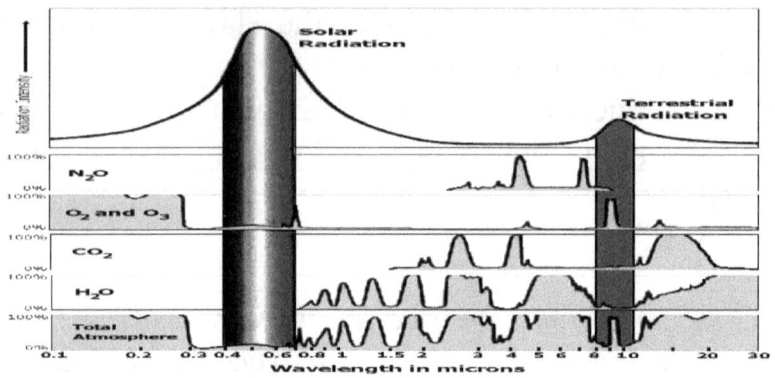

A crying shame-Some vain scientists have speculated that when the CO_2 absorption gets to 560 ppm there will be a substantial temperature rise that could be devastating. [Oops! I laughed a little writing that one.] A massive rise in CO_2 already noted has driven our temperature to the unbelievable increase of about 0.3 degrees making a further increase to 560ppm form the current 390 have a total increase of less than 0.5 degrees. We simply are not affected by CO_2 in the atmosphere at all. 560ppm (the dreaded doubling), temperatures should rise by *another* 0.2 to 0.5°C *ONLY*. IPCC [NATOs Global Warming gurus] estimate of 2.0 to 6.0°C, and this totally unfounded and without scientific merit.

Not much CO2 in the First Place-If CO_2 in the atmosphere has risen from about 290 to 390ppm in just over 100 years, which has been recognized, and if only 5% of all the greenhouse gases are man-made, then we can conclude that only 5% of the extra 100ppm could have been caused by mankind. This is only 5ppm over 100 years so we can say the following:

*The other change is due to the 95ppm from **naturally produced CO_2**.*

Even if CO2 had ANY affect almost none is by MAN-
Added to this, IPCC claimed that we will get 3-5°C more warming in the next 50 or so years. Mankind would only contribute 5 to 10% of that. That is only 0.2 to 0.5°C, even if we keep using coal. Just think about it man-used CO_2 makes up only 5% of the 5% of the 1% of the atmosphere or 0.0025%. The following chart represents one view that is much less than 5% for man uses and remember CO_2 makes up the tiniest fraction of the atmospheric Greenhouse gases, so divide that tiny piece of the pie down much farther and see if it can have ANY effect on weather.

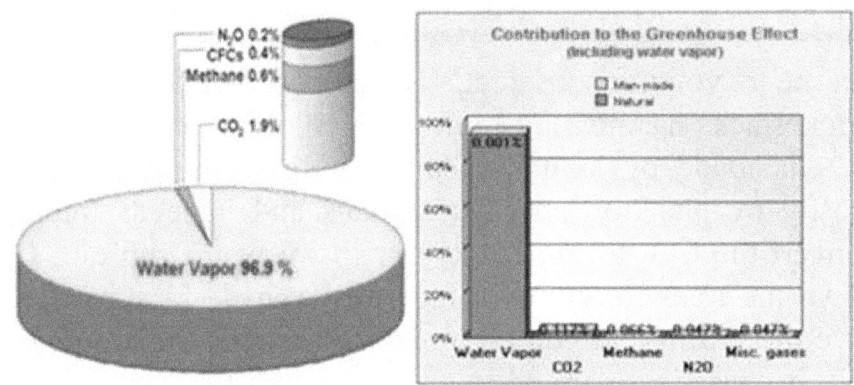

CO2 and Temperature

Massive temperature changes occurred well before man knew about coal- Let's hypothetically say CO2 is changing our temperature. The charts below show both CO2 levels and temperature captured in the Ice Cores from Greenland and Antarctica. The one on the left is from Antarctica over the last 50 thousand years. The thin line that begins below the erratic temperature curve shows something interesting. CO2 doesn't change until after temperature changes as temperature controls CO2 rather than the other way around. The second graph is from Greenland in case we didn't see things right, we see that Temperature, the erratic line changes well before CO2.

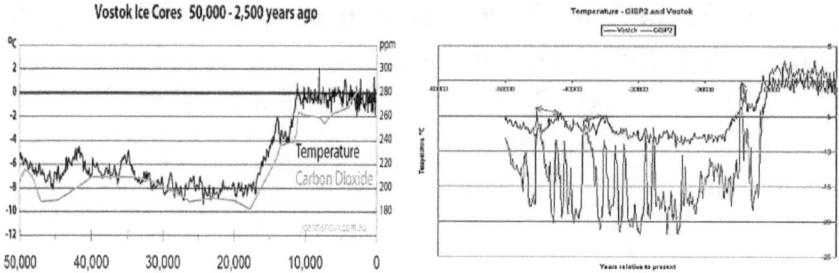

The second thing to notice is that the worse thermal change recorded in the last 15 thousand years happened 10000 years ago and very few automobiles were even on the roads.

Temperature increase makes CO_2 not the other way around

Let me let you in on a secret. The carbon dust coming out of smokestacks has nothing to do with colorless, odorless CO_2. CO_2 is made by animal respiration including people, but mostly by plankton and sea animals and it needs trees to convert the CO_2 to more Oxygen so we won't run out. Cut down the trees to allow biofuels to be made [a supposedly save the Earth] and you are going to increase the amount of CO_2 because there are no trees. This is what is happening in Europe.

Destruction because of the Lie-Because of all of the fear tactics of these vain scientists, today Europe has mowed down massive forests to produce biofuels so that they can say they are not killing the earth and without trees to absorb CO_2 they are increasing the issue and killing the Earth [but they seem to feel better because vain scientists are lying to them]. Bio-fuels from grain are not helping the situation. Besides the obvious loss of CO2 eaters [trees], use of "grain based fuels" will greatly increase food prices and roughly 30 million people are expected to be severely deprived. The USA will use up to 30% of the annual corn crop for alcohol production for vehicles alone. Ethanol production requires energy as well to make it economical which is generally done with those nasty hydrocarbon methods. The actual cost/gallon is much the same as other liquid fuels, but the miles per gallon consumed by vehicles are much lower than gasoline. One estimate is that one tank full of ethanol for an SUV is obtained from enough corn to feed one African for a year. Of course it isn't working by itself so worldwide ethanol plant subsidies in 2008 alone totaled more than $15 billion to reduce the food to feed the people of the planet to supposedly save them.

Ocean Boiling Lie-The huge increase in CO2 levels shown as a dotted line had absolutely NO effect on the Oceans that have stayed the same temperature. James Hansen's belief of CO_2 caused global warming is not supported by the tropic's data in the least nor is his crazy prediction of boiling oceans. You can keep driving you gasoline car and the earth will not even know it

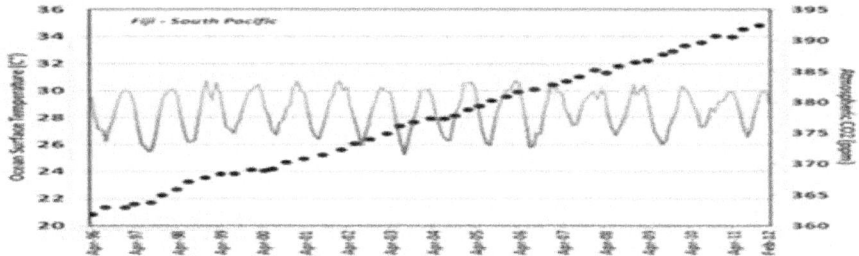

Temperature Trend Lie- If we expand out to the last 4000 years, the Greenland Ice core shows temperatures mostly stayed the same the whole time except for the little dip we started coming out of about 300 years ago. Since that time, the temperature has been trying to get back to NORMAL and the vain scientist KNOWS IT!

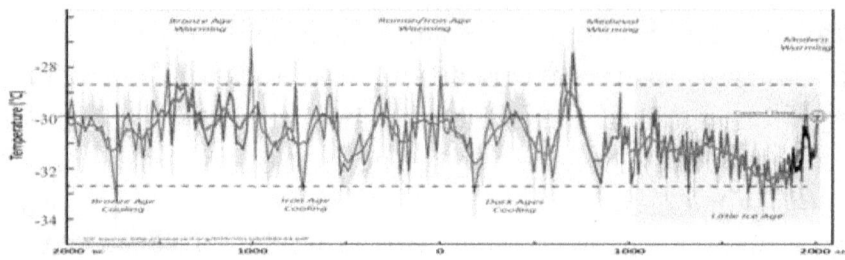

ARGOS Data Treachery-Scientists designed 3000 ARGO buoys that just floated around and took temperature measurements since 2003. These buoys show absolutely NO thermal increase [but the published data from NOAA somehow showed massive changes that the buoys "SOMEHOW" missed. The NOAA team decided that the information from them should not be used. One reason noted is that they weren't floating near the Arctic. Some might wonder why NOAA would have helped place these things and later decide they were stupid. Some begin to suspect that there was a large network of politicians, corporations, and scientists that were conspiring to promote the fear of "global warming" . . . despite evidence clearly stating no such "global warming" exists. With only $22 billion being pushed into the global warming epidemic, you might wonder why some would try such a scare tactic.

NOAA Published Truth then pulled it to publish a Lie- The National Climate Data Center and National Oceanographic and Atmospheric Administration [NOAA] put out a chart showing there was no significant temperature rise in the United States from 1940 until 2010 and that the spring was the coldest in the 115-year record, but at the same time they told everyone to ignore this data and focus on eliminating the Coal Industry.

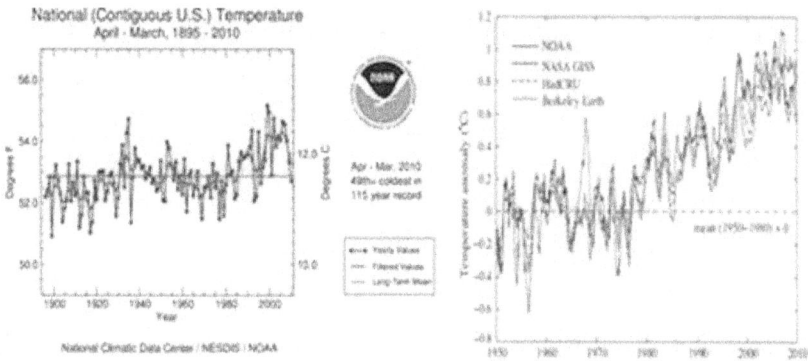

After realizing their mistake they put out a completely different chart so that people would fund their pet projects associated with Earth annialation. By focusing on the slight rise since about 1960, the danger begins to look real. We are going to look into this chart that has been the foundation for many, many others and is still being used even after the fraud was fully exposed last year. You will be seeing this same massive thermanl ramp again and again, but I will show you the true slope for the original data from the Bouys and sattelites.

17 Year Cooling Disregarding Lie-NOAA got fancy with this one. According to NASA's own data, the world has warmed 0.36 degrees Fahrenheit over the last 35 years, starting at the fairly cold year-1979. Even this would show a massive increase of 0.1 degree per year over that short time. The NASA Remote Sensing Systems data also shows that since 1998, the average temperatures around the world have been steadily decreasing as shown below.

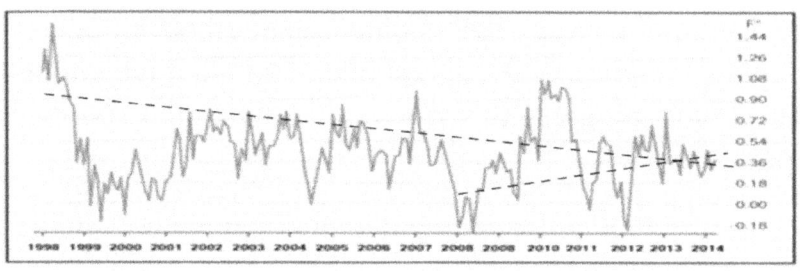

According to this graph, the world is 1.08 degrees cooler than it was in 1998. NOAA has not used this data when "Informing the World" about the condition of our Earth. To make it look like horror, just truncate until a rise is shown [2008 until today shows a 0.32-degree rise]. This is not the new data I mentioned to correct the previous longer-term graph. That is next.

Just Plain Lying-NOAAs current US graph is shown below left [same as before]. <u>*Now we know it is all a lie.*</u> Note that there is a discontinuity at 1998, which doesn't look right. <u>Globally, temperatures plummeted in 1999-2000</u>, but they didn't in the US graph. Note that measured data below right shows that by 2008, temperatures were back down to the 1989 level. But in the NCDC data, 2008 is half a degree warmer than 1989 making the temperature LOOK like a disaster when there is almost no change at all. Please note that the faked chart to the left has been used to justify an enormous number of "charts showing the destruction of our world.

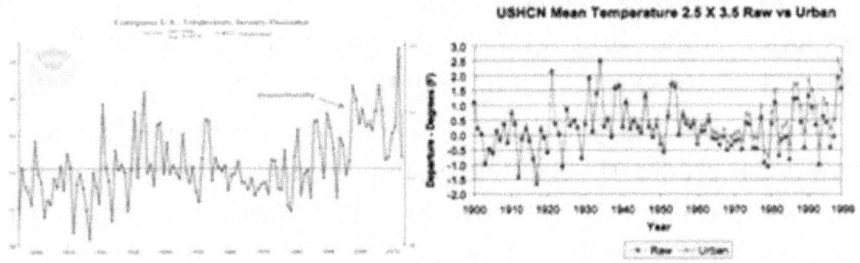

By putting the 2 togerther we can easily see the treachery. The top graph is from RAW data and the bottom one is the "doctored" chart making everyone want to give money to Green Technologies to protect them from this FAKE temperature rise.

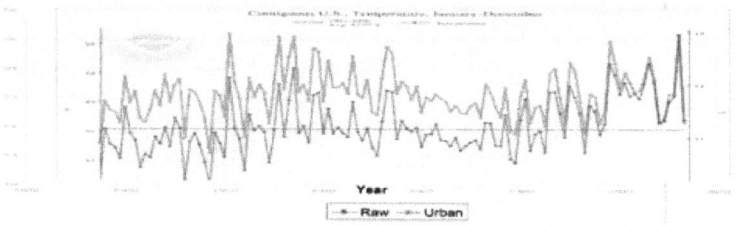

Top line is truth. Bottom line is the MANIPULATED NOAAs US temperature record is completely broken, and meaningless.

Adjustments that used to go flat after 1990 now go up exponentially. Adjustments which are documented as positive are implemented as negative to amplify fear for monetary gain.

Just erase the high temperatures of 1940-The best way to describe the subterfuge is to talk about the 1940s heat wave. This was a huge thermal "spike" occurring as the Earth recovered from the mini-Ice Age. If you were somewhat devious, you would like that spike to go away as it shows the temperature today is not significantly different than 1940 levels. This can be done 2 ways. Show the thermal rise since 1965. This is the favorite one, but you can also change the 1940 peaks to smooth out "anomalies that don't go along with the "THREAT". In this case it seems unthinkable, but this is another way to make people buy Solar Cells for their homes; especially when the US Government pays for most of the installations to eliminate the use of Coal that we have mountains of, is the lowest cost energy producer, and assures we can stay energy independent. The following shows the actual temperatures recorded and the one without the 1940s data is what you see on TV.

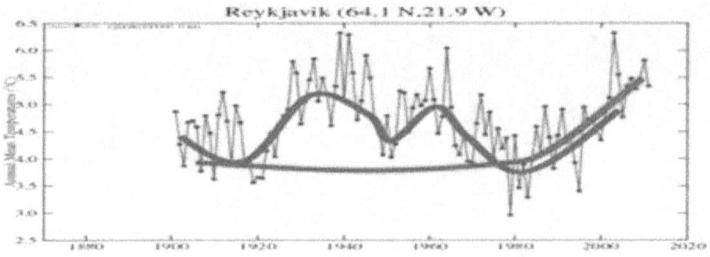

After the practice was revealed from some cleaver undercoverwork and retreval of internal EMAIL traffic, we have started to see a few of the practicioners coming clean about their part in the underhanded fear mongering.

"Ice" Lie-The Nobel Peace Prize, get rid of coal, billionaire stated the following, *"The North Polar ice cap is falling off a cliff. It could be completely gone in summer in as little as seven years. Seven years from now."* The images below show the Ice cap in 1990, 2007, 2013 and 2015. From the minor reduction in 2007, the Ice caps have been almost steadily increasing to over 63% larger.

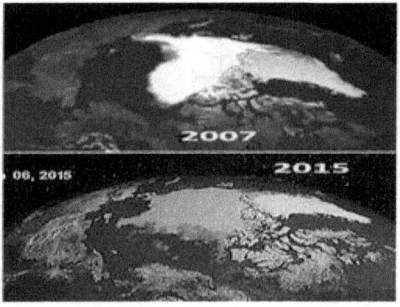

"Man Caused Death" Lie-Al Gore claimed CO_2 emissions *from Human factories were destroying our world as massive amounts of factory effluent were making the Earth's temperature go out of control and <u>97% of scientists agree it's real</u>, it's man-made, and it's dangerous.* Certainly, he knew the satellite data and all the rest, but he was making a fortune. He also knew many scientists were begging for people to listen to them as Mr. Gore misrepresented everything.

The Sun not CO_2-All those greenhouse gas things are the real reason for temperature fluctuations. The real culprit is the sun. All of the heat in our atmosphere is solar heat. If the sun gets brighter and burns away clouds, things happen and when the sun generates cosmic and X-Rays that hit the earth other things happen. Instead of just shining, the sun blast energy out in spurts. If we look at the correlation between temperature and total solar irradiance we see a much better relationship and we can begin to understand CO_2 is not the main player. In fact; the tiny CO2 gas levels could not possibly affect weather no matter how many gas guzzling cars there were! It is the activity of the sun (sun spots, solar flares, modification of other galactic cosmic radiation from outer space, the effects of solar wind, and magnetic flux), that affects the radiation arriving on earth. Here is a big one. The sun moderates cloud cover! Approximately 1% of the atmosphere is greenhouse gas and 90-95% of that is water. CO_2 is about 0.05% of the atmosphere. But only 5% of that 0.05% is man-made!

At 1.3 Billion times as large as the Earth, the sun makes 99.9999% of all the energy locked in our atmosphere. Some even suggest is gets colder at night and warmer in the daytime when the sun heats the air. From the next graph we can see that from 1978 the temperature is getting slightly warmer in parts of Greenland, so what did we do differently before 1978. The graph shows that there has been no appreciable slowdown in the increase of CO_2 in the atmosphere, but there was a fairly significant reduction in the temperature between 1940 and 1978. Sunspots control our heating if we want to limit heating, eliminating CO2 emission will have NO EFFECT.

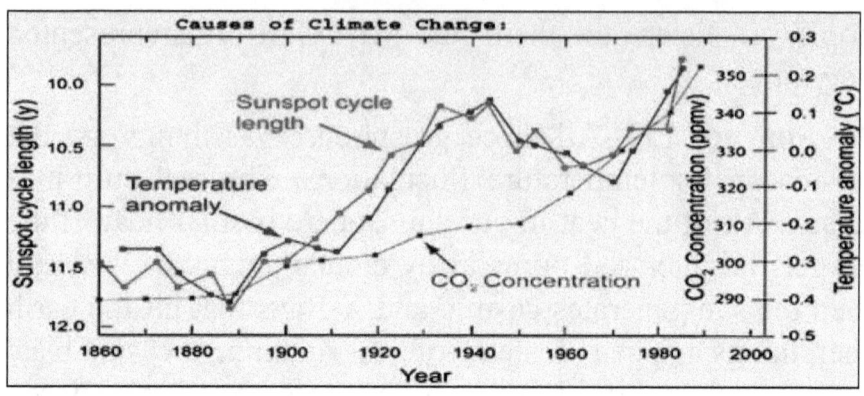

If you still think we are heading for a calamity and the only way to stop it is to keep cows from farting, shutting down jet aircraft using dirty jet fuel, and keeping forests from burning, please know China has increased its export of Coal processing plants for electricity and the percentage of extra costs for anti-greenhouse gas emission of cars is skyrocketing. Also, the use of corn ethanol instead of gasoline is destroying engines and our farming incentives which contributes to fewer jobs. The manufacture of electricity with solar cells and win vanes is showing to be environmentally damaging. Pilot whales beach themselves when they come near the horrible noise of the wind farms and thousands of birds are killed including Bald Eagles. Solar cell losses from weather and age are still major issues in a system that cost 10 times as much as coal----- and these industries are all built on lies. Speaking of lies, let's quickly talk about the flat Earth of Columbus.

About the Author

Steve Preston is a long-time author of scientific, esoteric facts. His series on the creation of mankind is shown below. The series focuses on the painful truths rather than whitewashed details that make us comfortable. If you are interested in the truth instead of comfort, please continue to read and, while you are at it, review other works by Mr. Preston as shown below.

Development of Mankind
The First Creation of Man-book 1 History of mankind
The Second Creation of Man-book 2 History of mankind
The Creation of Adam and Eve-book 3 History of mankind
The Antediluvian War Years-book 4 History of mankind
Man After The Flood-book 5 History of mankind
Close Look at Ancient History-book 6 History of mankind
A New View of Modern History-book 7 History of mankind
The Twentieth Century and Beyond- Book 8 History of Mankind

Bible History, Correction, and Analysis
Abraham to Moses-First part of the Bible
Adam's First Wife-Story of Lilith
Adam to Abraham- Second Part of the Bible
Closer Look At Genesis- 200 ancient text confirm Genesis
Exploring Exodus- Reviewing the Details of "Exodus"
Errors in Understanding- Interpretations of the Bible
Expanded Genesis- Apocrypha and other Jewish texts
Exploring Genesis- Reviewing the details of "Genesis'
Incarnations of God- How often did God become Incarnated?

History Confirmed By The Bible- Science confirms the Bible
Moses Saved Egypt- How the Jews eliminated the Hyksos
Moses to Jesus- Third part of the Bible Series
Mysteries of the Exodus- Proofs of the Exodus
New look at the Bible- Questions in Interpretation
Old Testament Used By Jesus- Ancient Jewish texts
Understanding the New Testament-4th part of the Bible Series
Why the King James Bible Failed- Issues with KJB

Anomalies

Religious Anomalies-Explaining unexplained scripture
US History Errors- Explaining disregarded events and causes
Planet Anomalies- Explaining anomalies of our near planets
DNA Anomalies- Explaining hidden truth tested by DNA

Ancient Technology and Life

Anakim Gods- History of the Ancient Giant/gods
Ancient History of Flying- Ancient flying
Kingdoms Before the Flood- Pleistocene humans
Living on Venus- Venus before the Pleistocene Extinction
Martians- Ancient Life on Mars
Mysterious Pyramids- Who made the Pyramids?
Victory of the Earth- History of our Earth
Not from Space- UFOs are not from space.
Amazing Technology- Descriptions of prehistoric capabilities

Ancient and Modern War

America's Civil War Lie- Truth about the Civil War years
Behind the Tower of Babel- Story of the Bharata War
Driven Underground- Fear in the Bharata War
Four Armageddons- The 4 major wars that destroyed mankind
Six Deaths of Man- Destructions of mankind
World War Before- The Pleistocene War
World War with Heaven- The Angel and Anak War
World War Zero-The Bharata War
When Giants Ruled the Earth- History of the Titan Giants
Sex Crazed Angels- What caused the Heaven War?

Current Events and Fears

Allah' God of the Moon- Terror of Muslims
American School Disaster- fear in our country
Can We Save America? - Fear in the USA
Scythians Conquer Ireland- A History of Ireland
Fast History of MILES Training- Laser based Army training
Great American Quiz- Unusual details of American History
Make Your Own Global Warming
Truth About Phoenicia- The Evidence -First in America
Monsters are Alive- Post Pleistocene Monsters
Promote the General Welfare- Fear in USA
Our Very Odd Presidents- President review
Terror of Global Warming- Fake issue uncovered
The Antichrist- Many demonic possessed rulers
The Bad Side of Lincoln- Negative side of a great man
The Devil- Of Demons and their master
Vampires among Us- How Demons and Vampires are similar
Humans on Display- Slavery and Human Zoos

New Look at Physics

Amazing Technology- Pleistocene Technology
Anthropic Reality- We control our Reality
Consensus Science- Fake Science
Complex Earth- Truth behind Earth's development
Is Time Travel Possible? Science of Time Travel
Retiming the Earth- Eliminate of Nuclear Decay Errors
Releasing Your Consciousness- Beyond our SELF
Slip Through a Wall- How to walk through solids
Our 12-Dimensional Universe- New science of our Universe
Mystery of Photons and Light- Science of Photons
Of Heaven and Hell- scientific descriptions
Meaning of Life and Light- Detains of New Science
Vibrational Matter- New Science of Quantum Fluctuations
Does Science Confirm the Bible? - Application of Physics

New Look at Biology

DNA of Our Ancestors- Tracing DNA of ancient man
God Didn't Make The Ape- New science on ape Evolution
Lizard People- Mutated People of the Bharata War
Creation and Death of Dinosaurs- Why Dinosaurs died

Races of Men- Tracing DNA of Humans
Tracing Cro-Magnon to Jesus-
Self, Soul, Spirit- Three components of Life
Self-Virtualization- New science of reality
True Happiness- Self Actualism and Beyond
Life Resonance- Unusual capabilities of men
Awaken the Departed- We can talk to the Dead
Biophotonics and Healing- How Photonics used in medicine

www.ingramcontent.com/pod-product-compliance
Lightning Source LLC
Chambersburg PA
CBHW050207230526
45470CB00001B/280